# Lecture Notes in Earth Sciences

45

Editors:
S. Bhattacharji, Brooklyn
G. M. Friedman, Brooklyn and Troy
H. J. Neugebauer, Bonn
A. Seilacher, Tuebingen

Albert Günter Herrmann    Bernhard Knipping

# Waste Disposal and Evaporites

## Contributions to Long-Term Safety

With 39 Figures and 21 Tables

Springer-Verlag
Berlin Heidelberg New York
London Paris Tokyo
Hong Kong Barcelona
Budapest

Authors

Albert Günter Herrmann
Bernhard Josef Knipping
Institut für Mineralogie und Mineralische Rohstoffe
Fachgebiet Salzlagerstätten und Untergrund-Deponien
Technische Universität Clausthal
Adolph-Roemer-Str. 2 A, W-3392 Clausthal-Zellerfeld, FRG

Translated by

Ralph B. Phillips
Paul-Gossen-Str. 109, W-8520 Erlangen, FRG

"For all Lecture Notes in Earth Sciences published till now please see final page of the book"

ISBN 3-540-56232-X Springer-Verlag Berlin Heidelberg New York
ISBN 0-387-56232-X Springer-Verlag New York Berlin Heidelberg

© Springer-Verlag Berlin Heidelberg 1993
Printed in Germany

Typesetting: Camera ready by author
32/3140-543210 - Printed on acid-free paper

**... perhaps one day cognition will come prior to experience...**

From »Wohin der Wind weht. Tschernobyl - zwei Jahre danach.«
German TV channel Nord 3, August 5, 1989.

# Preface

Some of the major ecological and social problems of the present and future are the production, treatment, and disposal of anthropogenic wastes. This is equally true for sparsely and densely populated industrial areas, including large countries in which sites for waste disposal would seem to be readily available. Especially nonradioactive hazardous wastes with their long-term toxicity need to be isolated from the biosphere just as effectively as radioactive substances. The long-term safety required of waste disposal sites can only be assured under specific geological and mineralogical conditions in certain parts of the lithosphere (underground repositories).

The subjects related to the production, avoidance, treatment, and disposal of anthropogenic wastes cover a range of knowledge encompassing the natural sciences, engineering, medicine, and law. This work presents some fundamental situations and problems concerning the disposal of toxic hazardous wastes which have been dealt with in several research projects.

The individual chapters are related scientifically. Long-term, effective solutions to our waste problems can only be found when interrelationships and possible future developments are considered.

Only the current status of this rapidly developing field can be discussed here. The individual chapters contain scientifically founded data and observations. Other aspects for which there are still controversial opinions and arguments are also discussed, which should stimulate further thought. Further developments and scientific advances can only be achieved by constantly challenging previous theories, and not through static observation and narrow-mindedness.

The most extensive quantification possible of the problems related to disposal of hazardous wastes is an essential aim of our work. This not only involves calculating the volume of waste and available repository space, but also compiling data on the long-term effects and the safe, long-term isolation of anthropogenic wastes from the biosphere. A simple description of conditions and processes without using concrete data, which is still widespread, is rejected since it frequently leads to pure speculation.

The scientific fundamentals and results presented in this work are of general validity for many questions concerning waste disposal. One example is the amount of waste produced annually in Germany, in which toxic, hazardous wastes play a major role. Following this train of thought, available data are used to show how limited the possibilities are for the long-term safe underground deposition of hazardous wastes with respect to the current quantities of waste.

Of utmost importance is information on the long-term effects of toxic wastes, as well as criteria which have to be considered with respect to the long-term safe deposition of hazardous waste. The natural chemical cycles and material transport in the various zones of the earth are the focus of interest here. They are the scientific basis for assessing every repository for anthropogenic wastes in geological systems. Therefore the significance of material transport and geochemical cycles is emphasized regarding all questions concerning the long-term safety of repositories on the earth's surface and in the lithosphere. Thus, our concept for the scientific evaluation of the long-term safety of underground repositories in geological systems differs from all other models presently under discussion in Germany.

In this work, marine evaporites are discussed with respect to the underground deposition of hazardous wastes and the long-term safety of underground repositories in salt rocks.

The isolation of hazardous materials from the biosphere can above all be influenced by fluid phases. Fluid phases can mobilize and transport hazardous materials through rocks in the biosphere. This is true, without exception, for all magmatic, sedimentary, and metamorphic rocks, and for marine evaporites, too!

In Germany evaporites have commonly been considered to be completely impermeable with respect to fluid phases (solutions and gases). This erroneous view stems from a complete lack of knowledge or misestimation of the dynamic evolution of the composition of evaporite bodies. Unfortunately, this is still true today for parts of some state agencies which deal with repositories. However, all observations of evaporite bodies made over the last more than 100 years have clearly shown that under certain conditions fluid and gaseous components are mobile in evaporites as well.

Solutions in marine evaporites have been the object of personal interest and scientific research of A.G. Herrmann for 40 years. The occurrence and formation of salt solutions in the various salt mining districts of Germany are presently being restudied and reevaluated on an extended scientific basis (e.g., v.BORSTEL 1992).

A presentation of the current knowledge on salt solutions is beyond the scope of this publication. However, in the interest of continuing research a research project proposed by A.G. Herrmann (1987b) will be introduced here. The direct quantitative analysis of the chemical composition (quaternary and quinary systems) of small fluid inclusions in rocks of the salt deposits of Hessen and Niedersachsen are the primary focus of this project. Information important to fundamental research on the formation and alteration of salt rocks and on the long-term safety of underground repositories should be gained from these studies (e.g., HERRMANN & v.BORSTEL 1991).

In addition to salt solutions, gases are also fluid components which occur in practically all marine evaporite deposits. Hence, both salt solutions and gases must be carefully considered when planning underground repositories in an evaporite body and evaluating their long-term safety. This publication contains an up-to-date overview of the gas occurrences in the marine evaporites of Central Europe. Despite previous studies, there is still a considerable deficit in scientific information regarding the distribution and formation of gases in the evaporites occurring in Germany. A detailed research program on the geochemical relationships involving the formation of evaporites and gases will draw attention to this situation.

One aspect must be emphasized in the planning and construction of repositories for anthropogenic wastes: their long-term safety. This publication deals precisely with this subject, and in Part III of this work we will present the concept that we have developed. This concept is based on the fact that evaporite bodies are subject to a dynamic evolution and that the chemical and mineralogical composition provides important information on the effect of fluid phases on salt rocks.

Previous works contain the testing of methods and presented initial results using the Gorleben salt dome as an example. However, we are just at the beginning of our research project on the long-term safety of underground repositories (e.g., HERRMANN & KNIPPING 1989, HERRMANN 1992).

The information contained in this publication is based on years of experience in evaporite research and underground repositories for anthropogenic wastes. Examples are presented which can be applied to similar situations and problems in other countries. Waste disposal is not just a national problem, it has long become an international one for all types of anthropogenic wastes.

Some of the results are from reports by »Der Rat von Sachverständigen für Umweltfragen« in Wiesbaden (HERRMANN 1991a), the Physikalisch-Technische Bundesanstalt, Abteilung SE, and the Bundesamt für Strahlenschutz in Salzgitter (HERRMANN 1988d, HERRMANN & KNIPPING 1989). We thank these institutions for their cooperation and the permission to publish the results in these Lecture Notes.

We would like to thank Mr. R.B. Philipps for translating this work. We must also thank Mr. J. Engelhardt, Mr. A. Schlegel and Mr. M. Siemann for their help in drawing some of the figures and in compiling the subject index.

Clausthal-Zellerfeld, February 1993

Albert Günter Herrmann
Bernhard Knipping

# Contents

Part I    Natural Geochemical Cycles and the Disposal of
          Anthropogenic Wastes

1       Subject ........................................................................................................ 3

2       Causal relationships ................................................................................... 4

3       What can be done? ...................................................................................... 5

4       The present waste situation ........................................................................ 6
4.1     Radioactive wastes ..................................................................................... 6
4.2     Nonradioactive wastes ............................................................................... 6

5       Long-term impact of nonradioactive and radioactive wastes .................... 12

6       Rates of migration and geochemical cycles on the earth .......................... 15

7       Possibilities for the construction of underground repositories .................. 19

8       Technologies for the mining of storage space in the underground ............. 22

9       Host rocks for underground repository sites .............................................. 30

10      Underground repositories in Germany ....................................................... 35

11      Limitations of underground repositories .................................................... 48

12      Assessment of the present situation ........................................................... 52


Part II   Fluids in Marine Evaporites

13      Salt solutions and fluid inclusions in marine evaporites ........................... 55
13.1    Scientific fundamentals .............................................................................. 55
13.2    Research concept ........................................................................................ 58

14      Gases in marine evaporites ........................................................................ 62
14.1    Fundamentals .............................................................................................. 62
14.2    Gas occurrences in marine evaporites ....................................................... 63
14.3    Inclusion of the gases ................................................................................. 65
14.4    Composition of the gases ........................................................................... 66
14.5    Gas volumes ............................................................................................... 72
14.6    Gas pressures ............................................................................................. 74
14.7    Formation and origin of the gases .............................................................. 75
14.7.1  Fundamentals .............................................................................................. 75
14.7.2  Hydrocarbons ............................................................................................. 77
14.7.3  Carbon dioxide ........................................................................................... 80

14.7.4   Hydrogen ........................................................................................... 80
14.7.5   Nitrogen ............................................................................................ 82
14.7.6   Hydrogen sulfides ............................................................................ 83
14.7.7   Noble gases ...................................................................................... 84
14.7.8   Other components ............................................................................ 84
14.8     What is the origin of the gases in evaporites? ................................. 84
14.9     Release of gases during mining operations ..................................... 89
14.10    Gas and isolated petroleum occurrences in the Zechstein evaporites of
         northern Germany ............................................................................ 90
14.11    Gases in the Gorleben salt dome ..................................................... 97
14.12    Adverse effects of salt-bound gases and radiolysis on the long-term
         safety of underground repositories ................................................ 102
14.13    Detection of gases and condensates in evaporites ......................... 105
14.14    Research concept ............................................................................ 107

Part III   The Composition of Salt Domes as a Criterion for
           Evaluating the Long-Term Safety of Underground
           Repositories for Anthropogenic Wastes

15       Geological principle ....................................................................... 113

16       The example of evaporites .............................................................. 115

17       Composition of Zechstein evaporites in the Hannover region, northern
         Germany .......................................................................................... 132
17.1     Evaporite rocks ............................................................................... 132
17.2     Fracture fillings .............................................................................. 137
17.3     Solutions ......................................................................................... 139
17.4     Gases .............................................................................................. 146

18       Calculation of mineral reactions and material transports ............... 148
18.1     Fundamentals for data processing ................................................. 148
18.2     Salt Rocks ....................................................................................... 152
18.2.1   Formation of kainitite from carnallitite ......................................... 152
18.2.2   Formation of K-Mg-free rocks from kainitite ............................... 155
18.2.3   Quantification of the reactions carnallitite →kainitite →K-Mg-free rock .. 157
18.3     Fracture-filling minerals ................................................................ 164
18.3.1   Crystallization of fracture-filling minerals ................................... 164
18.3.2   Quantification of halite crystallisation in cracks and fissures ....... 165

19       Evaluation of the current situation ................................................. 173

20       References ....................................................................................... 175

21       Subject index .................................................................................. 191

**Part I**


**Natural Geochemical Cycles and the Disposal of Anthropogenic Wastes**

# 1 Subject

Nature cannot be observed in isolated scenes from the past and present. The environment is the result of billions of years of dynamic evolution in the solar system and on the earth. The processes of evolution proceeded unaffected by man until the recent past. Now, the effects of the intervention of the *Homo sapiens* are becoming obvious. Hence, nature can no longer be understood based solely on the past. It is of existential importance that we begin to understand our environment in terms of future developments.

The past must be described, analyzed, and interpreted to be able to understand the present and predict possible developments in the future. This applies especially to all processes which formed the planet Earth over the last 4.5 billion years and which have changed in evolutionary and revolutionary ways in accordance with the laws of nature. The same or similar processes are also operating at the present, but at different rates - the principle of uniformitarianism. These processes will likewise affect the interior and exterior of our planet over the millions and billions of years to come. This knowledge is the scientific basis for the various disciplines in geosciences and this study.

Here, a field of work involving the long-term safe isolation of hazardous anthropogenic wastes from the biosphere will be presented. The significance of this work for the present and future will be evident from the onset. The scientific disciplines involved in this field are governed by the composition of the materials to be studied, their properties, and the changes caused by physical and chemical processes in dependency on time and space. Several causal relationships must first be pointed out to better understand the problem.

## 2 Causal relationships

For centuries, research in the natural sciences has been directed primarily at documenting the state of the environment and recognizing the laws of nature. The growing perception of the effects of human activities and manners of behavior on the environment is, however, one of the most recent and outstanding developments in the latter half of the 20th century. The sudden turn from quantity to quality in scientific research was evoked, among other things, by the population explosion, the accompanying increased use of raw materials and natural resources, and the resulting effects of solid, liquid, and gaseous anthropogenic pollutants on our environment. Much of the scientific knowledge we have acquired to date is of existential importance to the biosphere now and in the future. Thus, this knowledge has greater and greater consequences for practically all areas of human thought and behavior. This is reflected in the activities of various groups of society, the decision-makers, and the diverse information provided by the media.

Meanwhile the spectrum of problems facing the individual can hardly be comprehended. Yet, there are several issues of which we become aware of again and again because they are equally important to the individual and great portions of the Earth's population. Among these are forest decline, the devastation of rain forests, changes in the ozone layer, the eradication and extinction of animal and plant species, and the decimation of soils. All problems related to the pollution of the atmosphere, hydrosphere, and lithosphere and the disposal of anthropogenic wastes in our environment surely belong to these issues as well.

# 3 What can be done?

Regarding the aforementioned issues there is one central problem which relates directly or indirectly to all other problems. This problem involves the questions of what can be done about and where do we dispose of the ever increasing quantities of anthropogenic waste. These questions are acutely important not only to densely populated industrial countries with small surface areas, but also to countries which allow the importation of large amounts of waste and to the protection of the entire atmosphere and hydrosphere of the Earth. The complexity of this subject exceeds the possibilities of the producers, cities, and countries occupied with the disposal of these wastes. Activities such as research, planning, and realization measures related to anthropogenic wastes need to be intensified on an international level.

Answers to the questions of what can be done and where do we dispose of the ever increasing quantities of waste include measures which can be divided into two groups:
- the avoidance, reduction, and processing of wastes,
- the disposal of wastes.

There are neither medium- nor long-term alternatives for the three points in the first group. Under no circumstances is deposition alone a suitable means for conquering our waste problems in the long run because unlimited space for long-term, safe landfills or underground repositories is just not available for the great quantities of waste in small, densely populated countries. Present attempts and practices of disposing of wastes through export to other countries or dumping in international zones (e.g., the oceans) only shift the problem to other natural regions in an impermissible way without solving it. At the present, however, it must be remembered that we are just starting to develop and apply technologies for reducing the waste stream. This means that all anthropogenic wastes which still cannot be avoided or processed will have to be disposed of in coming decades as well. Scientists together with decision-makers and the public will have to work harder on concepts for the safe, long-term disposal of these wastes as well as all concentrated wastes created by the thermal processing of wastes.

# 4  The present waste situation

The present waste situation will be elucidated with some current data from a geoscientific point of view using Germany as an example. A distinction between radioactive and nonradioactive wastes will be made here.

Facts and numbers presented in part I of this work are valid for the Federal Republic of Germany in its borders prior to October 1990 (West Germany). Reliable data on the kind and amounts of wastes or on potential underground repositories do not yet exist for the new federal states (former GDR). It is obvious, however, that the waste disposal situation will be even more problematic if the amounts from the new federal states are included in the balance.

## 4.1  Radioactive wastes

In the year 2000 Germany in the borders prior to October 1990 with a predicted annual nuclear power generation of about 24 GW will produce approximately 14 000 m³ of radioactive waste with negligible heat output per year, in addition to several 100 m³ of conditioned waste with high heat output (BRENNECKE & SCHUMACHER 1990). In light of recent developments in nuclear technology these figures are probably too high, rather than too low.

Germany plans to dispose of all its radioactive wastes at depths of several hundred meters in the lithosphere.

## 4.2  Nonradioactive wastes

In 1984 the Federal Republic of Germany produced about 250 million metric tons of solid wastes, of which construction waste and excavated ground made up 45.2 %, industrial wastes 42.3 %, and household and similar commercial wastes 10.3 %. In addition, sewage sludge (dry matter) accounts for about 1.2 %, and hazardous industrial wastes for 1 % (Fig. 1; SPIES 1985, 1987).

In 1984 about 86 million metric tons of waste were disposed of by public plants in the following ways: 89.9 % in dumps, 8.7 % incinerated, 0.8 % composted, and the remaining 0.6 % by chemical-physical processing or in hazardous-waste dumps (Fig. 2; STATISTISCHES BUNDESAMT 1987).

Germany produces about 285 million metric tons of wastes annually (GOVERNMENT PRESS RELEASE 1989), of which hazardous wastes make up about 5 million metric tons. The greatest portion of these hazardous wastes was disposed of in surface dumps.

Among the types of waste mentioned above the toxicity and safe disposal of nonradioactive hazardous wastes pose a particular problem. The term hazardous waste has been used in different ways. In some federal states certain wastes have to be registered (hazardous wastes), in others they do not. Therefore, the data on the quanti-

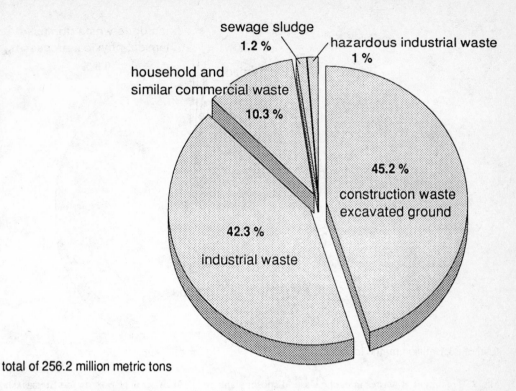

sewage sludge
1.2 %

hazardous industrial waste
1 %

household and
similar commercial waste

10.3 %

45.2 %

construction waste
excavated ground

42.3 %

industrial waste

total of 256.2 million metric tons

**Fig. 1** Quantities of solid wastes in the Federal Repulic of Germany for 1984. After SPIES (1985, 1987).

ties of hazardous wastes given by the federal and state governments differ. For 1983 the Federal Office for the Environment (Umweltbundesamt) reported that there were 4.9 million metric tons of hazardous wastes subject to registration (e.g., SUTTER 1987: 18; see also Tab. 1). These substances include sulfur-bearing wastes, combustion residues, lacquers, and paints.

According to the latest version and draft of the SONDERABFALL- UND RESTSTOFFBESTIM-MUNGS-VERORDNUNG (1989), the quantity of hazardous waste from the manufacturing industry and hospitals for 1984 was even 15.6 million metric tons. The sources of these 15.6 million metric tons of waste are given in Fig. 3 according to the branch of industry. The manner in which these hazardous wastes were treated and disposed of is shown in Fig. 4. Annually, 7.2 million metric tons (46.3 %) ended in surface dumps for hazardous waste, and 1 million metric tons (6.6 %) in underground repositories (several hundred meters deep).

There are prognoses of a reduction in the quantities of hazardous wastes by the year 2000. For example, the 4.9 million metric tons of hazardous wastes recorded by the Umweltbundesamt for 1983 should be able to be reduced by 50-60 % through avoidance and recycling (Tab. 1; SUTTER 1987: 26; see also HERRMANN 1988b, 1988c).

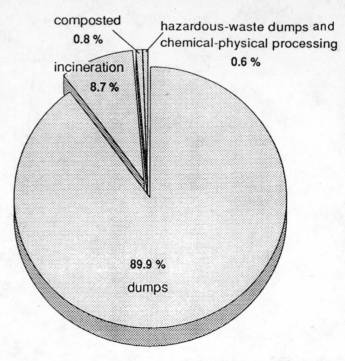

composted
0.8 %

hazardous-waste dumps and
chemical-physical processing
0.6 %

incineration
8.7 %

89.9 %
dumps

total of 86.1 million metric tons

**Fig. 2** Treatment of wastes in public waste-disposal plants for 1984. After the STATISTISCHES BUNDESAMT (1987).

In spite of this reduction, there would still be about 2.2 million metric tons of hazardous wastes subject to registration produced each year, of which about 1.5 million metric tons are solid and 0.7 million metric tons are liquid. The liquid hazardous wastes (e.g., petroleum products, organic solvents, paints) are to be processed - among other ways - thermally, primarily in high-temperature incineration plants.

Not only the data on the quantities of hazardous wastes but also on their manner of disposal differ from source to source and from publication to publication. In the federal state of Nordrhein-Westfalen, for example, around 165 000 t of hazardous waste was disposed of in underground repositories in 1984. For the same federal state 135 000 t is still projected for final storage in underground repositories for the year 2000. Extrapolating this figure for the entire Federal Republic of Germany in the borders prior to 1990 yields around 300 000 metric tons annually for final storage in underground repositories (MURL; see also HERRMANN 1988c). In all likelihood this latter figure is too low - rather than too high - when compared with the 1 million metric tons which has to be disposed of annually in underground repositories according to recent calculations based on the amount of hazardous waste produced in 1984 (SONDERABFALL- UND RESTSTOFFBESTIMMUNGS-VERORDNUNG 1989).

**Tab. 1** Quantities of hazardous waste subject to registration in the Federal Republic of Germany (in the borders prior to 1990) produced yearly, separated according to the waste type (from SUTTER 1987; supplemented by calculations of HERRMANN).

| No. | Types of waste | Year 1983 (solid and fluid) · 1 000 t | % | Year 2000 (solid and fluid) · 1 000 t | % | Year 2000 (solid) · 1 000 t |
|-----|----------------|------|------|------|------|------|
| 1 | sulfur-bearing wastes | 2 160 | 44.4 | 432 | 19.6 | 432 |
| 2 | oil-bearing wastes | 490 | 10.1 | 245 | 11.1 | - |
| 3 | combustion residues | 260 | 5.4 | 260 | 11.8 | 260 |
| 4 | paints, laquers | 250 | 5.1 | 75 | 3.4 | 75 |
| 5 | organic solvents (halogenous) | 233 | 4.8 | 46.6 | 2.1 | - |
| 6 | galvanic wastes | 192 | 4.0 | 58 | 2.6 | 58 |
| 7 | polluted soil | 166 | 3.4 | 166 | 7.5 | 166 |
| 8 | salt slags | 125 | 2.6 | - | - | - |
| 9 | filter media (kieselgur, activated carbon) | 108 | 2.2 | 108 | 4.9 | 108 |
| 10 | blast furnace sludges | 108 | 2.2 | 108 | 4.9 | 108 |
| 11 | organic solvents (nonhalogenous) | 90 | 1.9 | 27 | 1.2 | 15 |
| 12 | barium sulfate sludge | 45 | 0.9 | 45 | 2.1 | 45 |
| 13 | other (estimated) | 633 | 13.0 | 633 | 28.8 | 300 |
|  | Total | 4 860 | 100 | 2 203.6 | 100 | 1 567 |

However, the figure of about 1 million metric tons was recently given alone for hazardous waste that is supposedly only suitable for final storage in salt caverns, i.e., a special type of underground repository (GOVERNMENT PRESS RELEASE 1989).

More on the subject of hazardous waste considered for underground disposal is given, for example, by WIEDEMANN (1988b), but no data on the quantities of individual types of waste produced annually are provided.

One thing is evident regardless of the calculations of the quantities of hazardous wastes being produced: *There will be 100 to 1000 times more solid, highly toxic, hazardous wastes and/or substances with components that are readily mobilized, such as ashes, slags, sludges, metallic wastes, salts, and different types of hazardous wastes* (see SONDERABFALLARTEN-KATALOG 1989; SUTTER 1987; HERRMANN 1988b, 1988c), *than there will be radioactive wastes for many years to come. And the quantities of nonradioactive hazardous wastes which have to be disposed of annually in under-*

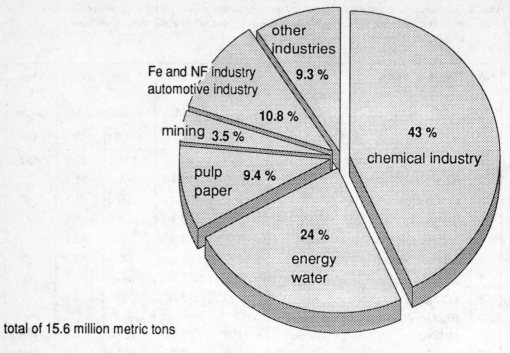

total of 15.6 million metric tons

**Fig. 3** Source of hazardous waste for 1984 according to branch of industry. After the SONDERABFALL- UND RESTSTOFFBESTIMMUNGS-VERORDNUNG (1989).

*ground repositories for the purpose of long-term safe isolation from the biosphere are 10 to 100 times greater than those of radioactive wastes.*

Hence, the question must be asked whether the toxicity and the long-term effect of nonradioactive wastes and contaminants are to be classified in the same way as those of radioactive substances.

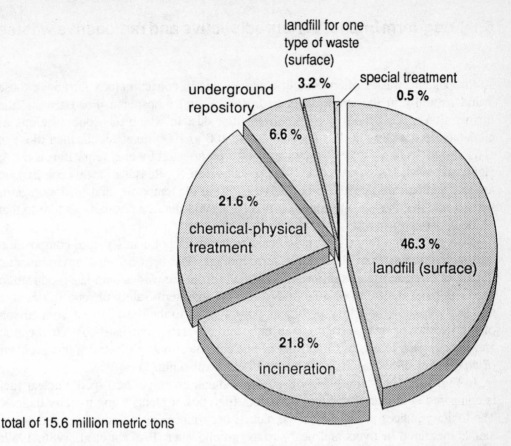

total of 15.6 million metric tons

**Fig. 4** Treatment and disposal of the hazardous wastes recorded for 1984. After the SONDERABFALL- UND RESTSTOFFBESTIMMUNGS-VERORDNUNG (1989).

# 5 Long-term impact of nonradioactive and radioactive wastes

Anthropogenic wastes often contain contaminants in concentrations far above those found naturally in the atmosphere, hydrosphere, and lithosphere. For example, the incineration of household trash produces filter dust in which the concentrations of elements such as Bi, Cd, Pb, Sn, and Zn are 100 to 1000 times higher than those in soils and rocks (e.g., shale) which have not been affected by man (e.g., BRUMSACK & HEINRICHS 1984; HERRMANN et al. 1985; see also Tab. 2). Reasons for this comparison are that geological systems are our only option for the temporary and final storage of wastes, and that the average composition of soils and shales are usually similar to that of the upper earth's crust.

Now and in the future man must learn to live with pollutants whose composition are very different from the natural surroundings. The behavior and environmental impact of these substances differ fundamentally from the wastes man has produced in centuries past. Hence, to be able to evaluate the long-term safety of repositories it is necessary to know how long pollutants pose a threat to the biosphere. A comparison of the toxicity of nonradioactive and radioactive wastes from hard-coal combustion and the radioactive wastes of the fuel of nuclear power plants illustrates this problem (EHRLICH et al. 1986, 1987; RÖTHEMEYER 1988; see also Fig. 5).

In Fig. 5 the long-term effect of reprocessed and unreprocessed spent nuclear fuel is compared with that of residues from coal-fired power plants using toxicity indices. The toxicity indices are related to the natural occurrences of radionuclides and heavy metals measured in rivers and near-surface groundwater (EHRLICH et al. 1986, 1987; RÖTHEMEYER 1988). Fig. 5 shows that the toxicity of radioactive wastes from nuclear power plants decreases with time due to the radioactive decay of the radionuclides. In contrast, the toxicity of nonradioactive wastes remains nearly unchanged for, in part, very long periods of time (the residues of coal-fired power plants are only used here as a representative example). The toxicity of wastes from reprocessed nuclear fuel, for instance, falls below the nearly unchanged toxicity level of nonradioactive filter dusts and ashes from coal-fired power plants after about 400 years. The toxicity of unreprocessed spent fuel (direct disposal after use) drops below that of residues from hardcoal combustion after about 2000 years (Fig. 5; EHRLICH et al. 1986, 1987; RÖTHEMEYER 1988).

To avoid any misunderstandings it is emphasized that in Fig. 5 the specific properties of radioactive wastes are not being equated with those of nonradioactive wastes. A statement on the advantages and disadvantages of coal-fired and nuclear power plants is also not intended. This comparison should only point out a fact which is of central importance to the long-term safe disposal of wastes. This can be summarized as follows:

*With respect to their long-term impact and thus the long-term safety of their disposal equal consideration must be given to the annually 100 to 1000 times greater*

**Tab.2** Average chemical composition of filter dust from hard-coal-fired plants, municipal refuse-fired power plants, and municipal sewage sludge in µg element/g waste (= ppm). The values in ( ) indicate the relative enrichment of metals in the filter dusts and sewage sludges compared with the mean natural and unaffected by man composition of soils and shales (BRUMSACK et al. 1984; HEINRICHS et al. 1984; BRUMSACK & HEINRICHS 1984; HERRMANN et al. 1985)

| Element | Filter dust, Hard coal | | | | Filter dust, Refuse incineration | | Municipal sewage sludge, average of 9 cities | |
| | Fluidized-bed combustion | | Refuse furnace | | | | | |
|---------|------|-------|------|--------|--------|---------|--------|--------|
| As | 447 | (60) | 155 | (21) | 59 | (8) | 5.4 | (1) |
| Bi | 5.4 | (42) | 1.8 | (14) | 25 | (192) | 5 | (38) |
| Cd | 34 | (262) | 21 | (162) | 184 | (1415) | 12 | (92) |
| Cr | 233 | (3) | 182 | (2) | 589 | (7) | 215 | (2) |
| Cu | 442 | (11) | 238 | (6) | 882 | (23) | 610 | (16) |
| Hg | 0.2 | (2) | 0.33 | (3) | 1.2 | (12) | 8.8 | (88) |
| Ni | 272 | (4) | 348 | (5) | 177 | (3) | 200 | (3) |
| Pb | 966 | (44) | 309 | (14) | 650 | (297) | 290 | (13) |
| Sb | 43 | (43) | 12 | (12) | 207 | (207) | 80 | (80) |
| Se | 36 | (360) | 19 | (190) | 11 | (110) | 1.3 | (13) |
| Sn | 18 | (3) | 38 | (6) | 2070 | (345) | 27 | (5) |
| Tl | 29 | (43) | 6.8 | (10) | 2.1 | (3) | 0.2 | (1) |
| Zn | 1 400 | (12) | 756 | (7) | 17 300 | (150) | 2 100 | (18) |

*quantities of nonradioactive wastes, and the very small volumes of radioactive substances.*

The consequences of this have not yet been fully recognized either by lawmakers or by the public. Therefore, the necessity of testing the long-term safety of repositories for nonradioactive wastes with the same prudence as that for radioactive wastes of all types must be stressed again and again.

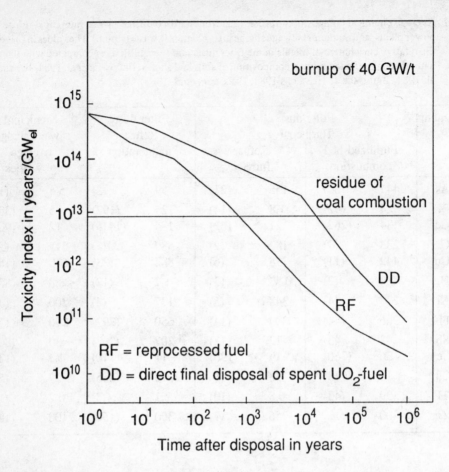

**Fig. 5** The long-term impact of nonradioactive and radioactive residues from coal-fired and nuclear power plants. The toxicity of the wastes is given in relation to one GW of generated electric energy. The calculations are based on measurements of the natural radioactivity and amounts of heavy metals in rivers and near-surface groundwaters (from RÖTHEMEYER 1988; see also EHRLICH et al. 1986, 1987).

# 6 Rates of migration and geochemical cycles on the earth

How and where can anthropogenic wastes, especially hazardous wastes, be disposed of safely for long periods of time? Which factors must be taken into account? What are the limitations of the measures for the long-term safe disposal of wastes? In this chapter some of the fundamental answers to these questions will be discussed from a geoscientific point of view.

*Wastes can only be disposed of in geological environments at the earth's surface (landfills) and/or in the underground (mines).* This statement is of general validity and fundamental significance, and applies equally to all types and quantities of anthropogenic wastes.

For various reasons it will not be possible either at present or in the near future to dispose of anthropogenic wastes in space. Because of the risk connected with the lift-off of spacecraft, the transport of containers bearing radioactive or nonradioactive contaminant concentrates is out of the question. Furthermore, the quantities of nonradioactive wastes are so great that their transport into space simply would not be financially feasible. Last but not least, it must be remembered that space and our solar system will be the object of scientific study in the next millenium.

Since the disposal of anthropogenic wastes is limited to geological environments, the geosciences (geology, mineralogy, geochemistry, geophysics, and geography) will play a major role in determining the steps to be taken for the safe long-term disposal of these wastes on the earth.

There are two facts which are of fundamental significance to all repositories for anthropogenic wastes:

1. *The earth is not in a static, but in a very dynamic state. This means that migration and geochemical cycles exist within and between the atmosphere, hydrosphere, and lithosphere. Hence, in the future all wastes buried in the earth will also participate in these natural processes, which cannot be influenced by man. This fact is the focus of all repository projects for anthropogenic wastes both at the surface (landfills) and underground (mines).*

2. *Long-term observations and experiments are needed to prognosticate the long-term safety of repositories (especially underground repositories). The natural environment is the only laboratory for such studies. Both rapidly and slowly progressing mineral reactions and geochemical migration are documented (preserved) here. In contrast, only short-term experiments are able to be conducted in the laboratory, and the results extrapolated for longer periods of time. Our knowledge of long-term safety from the laboratory must hence be supplemented by observations of nature.*

The following observations of nature have provided information for the planning and realization of waste repositories at the surface and underground:

The migration, interaction, and mixing of solid, liquid, and gaseous components occur much quicker at the earth's surface, in the atmosphere, and in the hydrosphere than in the lithosphere at depths of several hundred and thousand meters. An impression of the duration of such geochemical cycles in the various environments of the earth is given in Fig. 6. The durations shown are generalized: they may be shorter or longer depending upon the specific element.

In the atmosphere the residence time of many components is on the order of years. In the hydrosphere and pedosphere the residence times of hundreds and thousands of years are estimated. There is a great difference in residence times between these three environments at the earth's surface and rocks deeper in the lithosphere:

*Depending on depth and rock type the geochemical cycles in the lithosphere take considerably longer than at the surface, i.e., millions of years.*

## Conclusion

Landfills are always in an environment of comparatively rapid geochemical migration, even though climate also influences speed of migration. Consequently, at the earth's surface a long-term effective isolation of contaminants from the biosphere cannot be assured in spite of all preparations and conceivable engineered structures (e.g., IWS-Schriftenreihe 1, 1987; RÖTHEMEYER 1991: 37ff).

Accumulating reports on the threat to groundwaters and soils posed by industrial dumps after just a few years or decades can be explained by natural geochemical migration and cycles near the earth's surface. In West Germany, for instance, the 95 known hazardous-waste dumps annually release 1-3 million m³ of seepage, which is extremely difficult to retain and treat (GOVERNMENT PRESS RELEASE 1989). In view of the geochemical cycles in the environment at the earth's surface discussed above it must be presumed that cleaning up these surface dumps, which would cost up into the billions, would only have a short-term effect on safety. Thus, such measures are only a temporary solution to these problems, which will have to be rectified again in the future at great expense. Long-term safety for future generations cannot be achieved in this way.

In this context WIEDEMANN (1988a: 936) must be mentioned. He proposes that the unhindered erosion of exposed surface dumps (waste banks) by water, wind, and frost can easily be recognized as well as combated with simple measures. This clearly shows that surface dumps simply do not offer a long-term safe isolation of contaminants from the biosphere. In addition, the assumption that the mobilization of contaminants in surface dumps can be stopped easily with simple methods must be rejected. The problems, especially the financial ones, presented by existing industrial dumps clearly prove the opposite.

Based on the previous remarks on natural geochemical migration and cycles it is understandable that the long-term isolation of contaminants from the biosphere and thus the long-term safety of waste repositories can only be achieved at depths of

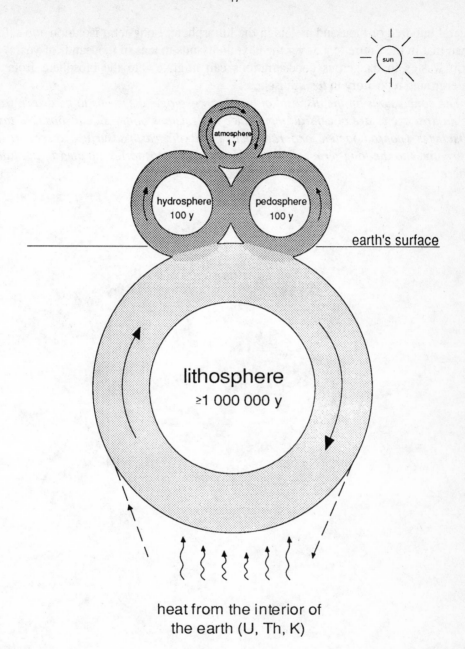

**Fig. 6** Durations of geochemical cycles and residence times of elements and compounds in and between the atmosphere, hydrosphere, pedosphere, and lithosphere. After an idea of GOLUBIĆ et al. (1979: 37). Residence times after HERRMANN (1988b, 1988c).

several hundred or thousand meters in the lithosphere. Long-term isolation and safety mean that in the future (e.g., over the next thousands to tens of thousands of years) no toxic wastes in dangerous concentrations can migrate into the biosphere from an underground repository in the lithosphere.

*The conclusions on the disposal of wastes in geological systems to be drawn from the natural cycles are of central importance to decisions on the safe disposal of toxic substances (nonradioactive and radioactive). At the earth's surface there are no alternatives to the long-term safety of underground repositories situated in the lithosphere.*

# 7 Possibilities for the construction of underground repositories

Various types of underground repositories are being discussed, some of which are already in use. Repositories can be established in mines and as caverns at depths of several hundred or thousand meters (Fig. 7). Caverns can be created in salt domes (below the water table) by selectively dissolving rock salt with water (Fig. 8). Another possibility is the mining of horizontal shafts and caverns in mountainous areas with caverns (above water table; Fig. 7). Silicate rocks such as granite and crystalline shales are preferred hosts for such caverns. Finally, drill holes several thousand meters deep are also being considered for the disposal of relatively small amounts of high-

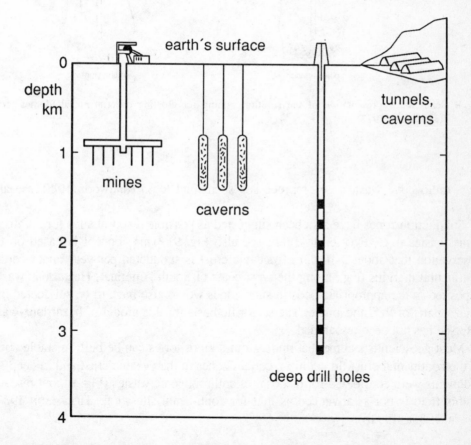

**Fig. 7** Types of geological repositories for the long-term safe final disposal of anthropogenic wastes in the underground (from HERRMANN 1987a).

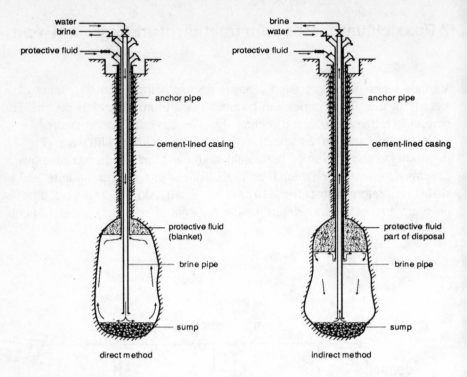

water —
brine —
protective fluid —

anchor pipe

cement-lined casing

protective fluid
(blanket)

brine pipe

sump

direct method

brine —
water —
protective fluid —

anchor pipe

cement-lined casing

protective fluid
part of disposal

brine pipe

sump

indirect method

**Fig. 8** Schematic representation of the solution mining for creating caverns in salt domes (from RISCHMÜLLER 1972).

level radioactive wastes, i.e., reprocessed spent fuel (e.g., RINGWOOD 1980; see also Fig. 7).

Subduction zones have also been suggested as possible disposal sites for anthropogenic wastes (e.g., FYFE et al. 1984; see also Fig. 9). This concept is based on the observation that about 2-10 cm of oceanic crust is subducted per year under certain continental margins (e.g., along the west coast of South America). Hazardous wastes deposited in the appropriate beds in such zones would also have to be subducted into deeper parts of the lithosphere. The scientific basis for this model of hazardous-waste disposal has not been researched.

Most geoscientists agree that underground repositories can be built in stable zones of the continental crust. In addition, certain zones in the oceanic crust and in deep-sea sediments were considered for disposal of anthropogenic wastes (Fig. 9), but research in this field is not as advanced as that for continental sites (e.g., HERRMANN 1983a, 1987a, 1988a, 1991b).

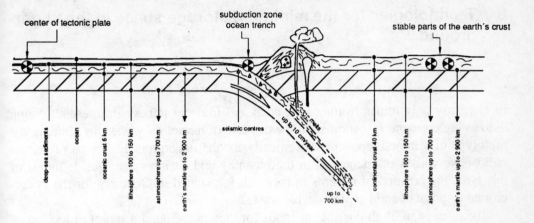

**Fig. 9** Zones of continental and oceanic crust considered for disposal of anthropogenic wastes (not to scale; HERRMANN 1987a).

Radioactive and nonradioactive hazardous wastes have been dumped in the oceans for decades. Various studies have shown, however, that ocean currents can distribute contaminants over large areas (e.g., KAUTSKY 1982). Consequently, the disposal of solid and fluid wastes in the oceans must be stopped.

# 8 Technologies for the mining of storage space in the underground

In Germany only mines (former salt, coal, and iron ore mines) are presently being used as underground repositories for the disposal of hazardous wastes. In addition, the prototype of a mined repository specifically for the underground disposal of certain radioactive wastes (Gorleben) is in the planning and design stage. Beside mines of this type, the creation of caverns in rock salt is planned in Germany for the underground disposal of solid anthropogenic wastes.

The discussion of all possible methods for the underground disposal of hazardous wastes will inevitably be revived in Germany in the near future. Hence, a short overview of the advantages and disadvantages of the various technologies and their feasibility in Germany will be given in the following.

## Landfills

Landfills are among the simplest methods for disposing of wastes. This form of disposal is used, for example, in the USA for radioactive wastes (e.g., MEYER 1976; ERDA 1977a: III-85 to III-93, 1977b: II-116 to II-126; ROBERTSON 1980). France is also planning to use this method of disposal for low-level radioactive wastes (BARDET 1976).

Landfills are several meters deep and wide and are lined with cement, asphalt or tar. Wastes are packed in steel or concrete canisters. The excavations are then backfilled with clay.

Experience has shown that this method of radioactive waste disposal does not assure long-term reliable isolation. Water does eventually penetrate the clay cover, and the canisters are not resistent to corrosion, allowing contaminants to leak out of the canisters and into the ground around and below the landfill. Accidents involving radioactive wastes stored in such landfills are good examples for the effectiveness of the rapid geochemical migration and cycles at and near the earth's surface (see ROBERTSON 1980; HERRMANN 1983a: 120).

Based on previous observations of geochemical migration in nature and previous experience with such sites an urgent warning against disposing of anthropogenic wastes in near-surface soil or rock must be given. The sealing of landfill with clays or other materials as recommended and practiced again and again by engineers cannot assure the long-term isolation of contaminants from the biosphere. Long-term safety here means thousands of years and longer, not just decades or centuries.

Underground disposal involves the storage of wastes at depths of at least several hundred meters in suitable repositories and rocks. The disposal of radioactive and nonradioactive wastes in landfills does not meet this requirement.

## Deep drill holes

This concept involves the sinking of 3000 to 4000-m-deep drill holes in bodies of silicate rocks or evaporites for disposal of toxic wastes in canisters. In the past this technique has been discussed exclusively for relatively small volumes of high-level radioactive wastes with high heat emission (e.g., MANAGEMENT OF COMMERCIALLY GENE-RATED RADIOACTIVE WASTE 1, 1979: 3.3; RINGWOOD 1978: 50; HOFRICHTER 1980a: 427; ELSAM & ELKRAFT Report IV, 1981: 50-69; HERRMANN 1983a: 120 ff).

A drill hole repository could be designed as a widely spaced grid in a body of rock of uniform petrographic composition. Such sites should be located in sparcely popula-ted areas with little biological activity and no significance to water or raw material supply. Deep drill holes have the advantage that they are sealed after being filled and require only a few years of work. New holes can be drilled as needed.

The drill holes must be sealed securely against any possible water- or fluid-bearing formations and the surface.

Drill holes sunk in salts could be sealed at the surface with, for example, cement, salt slack, and asphalt. Sealing the sides of the drill holes against solutions from deeper rock formations is more difficult. The geological estimation of such solution flux is complicated due to the lacking possibility for direct inspection. Indirect me-thods of measurement (geophysics) provide valuable information, but do not replace the direct geological study possible in mines.

Future developments will determine whether and to what extent the deep drill hole technology can be used for disposing of anthropogenic wastes. In any case, this technology will only be economically and technically feasible for small quantities of certain hazardous wastes.

In Germany there are many geological arguments against the widespread use of this technique for disposing of hazardous wastes, in addition to the lack of sparly populated areas. Silicate rocks can hardly be considered in Germany due to the lack of such large, compositionally homogeneous bodies of rock for the drill hole repository. The suitability of evaporites for this purpose is very limited as well. The Zechstein salt domes in northern Germany frequently contain carnallitic rocks in complicated folds. Carnallite is a very reactive mineral in the presence of aqueous solutions and temperatures above 100 °C. Therefore, a hazardous-waste repository must have a sufficient safe distance to carnallitic rock to ensure its long-term safety. Exploration drillings alone are inadequate to reconstruct reliably the spatial distribution of carnal-litic rock in a salt dome. This is only possible through mining.

In the Rotliegend-evaporites of northern Germany, however, the deep-drilling tech-nique might be able to be employed since they occur at several thousand meters depth, consist predominantly of rock salt and mudstones, and contain no carnallite-bearing rocks (HOFRICHTER 1980a: 427). Yet here as well, only small volumes of wastes could be disposed of. In Germany the deep-drilling technique will surely not play a medium- or long-term role in the safe isolation of larger quantities of hazardous anthropogenic wastes from the biosphere.

## Caverns

Large individual voids in bodies of rock are referred to as caverns. They are currently used above all for the underground storage of crude oil, liquid and gaseous chemical compounds, natural gas, and compressed air. The final disposal of certain solid wastes in such caverns is also planned. Voids can be produced by mining or dissolving of rock salt (caverns dissolved in the rock salt of evaporites; e.g., KNISSEL & FORNEFELD 1988; LUX 1988). Storage space can be created from the surface at depths of several tens of meters (e.g., the mined repository at Forsmark in Sweden for low-level radioactive wastes 50 m below the floor of the Bay of Bothnia) down to about a thousand meters (e.g., salt caverns). Another possibility involves constructing mined repositories with several hundred meters of cover through horizontal shafts driven in the base of the mountains, possibly combined with a system of shafts (Fig. 7). This concept is presently being investigated in Switzerland for the disposal of low- and medium-level radioactive wastes (e.g., NAGRA 1987).

In addition to the storage of liquid and gaseous raw materials in caverns dissolved in salt, Germany is also planning the ultimate, safe disposal of hazardous anthropogenic wastes in such caverns. The Jemgum salt dome in Niedersachsen will be the object of study in a project on dissolving caverns in salt (Chapter 10, Fig. 14). Studies on the disposal of radioactive wastes in caverns were conducted several years ago in the Asse test mine and repository. In this case, however, a small cavern was established from within the mine. Hence, this project and the Jemgum salt dome project, which is to be conducted from the surface, cannot be compared directly.

The mining of caverns and special underground repositories produces rock wastes which accumulate in heaps at the surface. Such rock is only rarely suitable for building purposes. This is not a problem in the case of salt caverns created by solution mining since the rock is dissolved - assuming the salt dome is situated near the coast (see below).

To produce caverns in salt by the solution mining a drill hole is first sunk into the salt dome from the surface. Water is then pumped through the hole until a bulb-shaped cavern forms (Fig. 8). In coastal areas seawater is used for dissolving the salt, which is then fed back into the ocean as a nearly saturated brine. The ocean is not polluted in this way because the dissolved salts crystallized out of seawater in the geological past. The rock salt (Rotliegend and Zechstein) underlying central and northern Germany, for instance, were formed by evaporation of seawater. However, when such saturated brines and the suspended matter contained therein are fed back into the ocean, any adverse environmental effects at the point of discharge or downstream until the brine is diluted must be examined. This is a controversial subject (e.g., HERRMANN et al. 1990).

The proposed volume of such salt caverns for disposing of hazardous wastes ranges from 75 000 to 150 000 m³ per cavern (about 300 m high and 45 m in diameter). The caverns would be situated at depths between 800 m (roof) and 1800 m (base). Solids (e.g., chemical salts), filter dust, slags, and sludges are to be disposed of in such caverns. Originally, caverns were considered as a repository for conditioned wastes. However, now the packing of wastes in drums before their final disposal in salt caverns is no longer considered (Fig. 10; ANSORGE 1987).

Hazardous wastes are to be piped directly from the surface into the caverns produced by solution mining. Consequently, no ventilation or special transport system is necessary in the caverns.

The situation is different for establishing such salt caverns in evaporites situated inland. In this case, the brines produced by solution mining can only be fed into rivers or stored underground in inactive salt mines. Certain conditions are imposed on letting saturated brines into rivers, and only rarely are these brines stored in inactive salt mines. Examples of the latter case are the two caverns in the Benthe salt dome southwest of Hannover for the natural gas supply for the city of Hannover; the brines from these caverns are being stored in the inactive Hope salt mine (Hope salt dome).

Several factors must be taken into consideration when using salt caverns produced with the solution method.

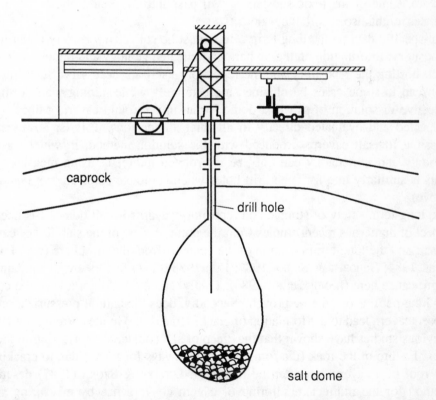

caprock

drill hole

salt dome

**Fig. 10** Example of a cavern in salt for the disposal of nonradioactive wastes (not to scale; after KBB-Untertagespeicher 1984).

1. Rock salt is not the only rock in a salt dome that can be dissolved for producing a cavern. Potash salt seams, which dissolve more easily due to the high solubility of K- and K-Mg minerals (e.g., sylvite and carnallite) may also be interbedded in the rock salt. In this way, pathways for solutions - or at least local horizons - into neighboring salt caverns can form. Such observations were made during the creation of caverns in the Etzel salt dome (used for storing federal crude oil reserves). It is evident that problems regarding the long-term safety of a cavern repository can arise due to potash salts and their high solubility. This is also true in a similar way for jointed anhydrite horizons which extend through cavern walls. Anhydrite beds are also preferred pathways for solutions in evaporites. There is a difference between salt caverns in which only liquids or gases are to be stored for a number of decades or hazardous-waste repositories whose safety must be assured for thousands of years. The liquids and gases presently stored in caverns can be retrieved, whenever necessary. The retrieval of solids from such a cavern, however, is practically impossible. The storage of hazardous wastes in salt caverns is final. Mistakes made while constructing this type of underground repository cannot be corrected.

2. When dissolving rock salt with water to produce caverns, some water does remain at the bottom of the cavern. This means that when solid wastes are deposited in the cavern, some of the toxic substances will pass into the liquid phase. These substances might also absorb the water.

3. Site-specific data on the long-term safety must be collected for every underground repository. Information on the, in part, complicated geology and mineralogy of the rock hosting the repository and the repository itself are necessary. Such information can in some cases be obtained more reliably while mining than during the selective dissolution of salt in a dome for an underground cavern. Mines can be inspected and evaluated directly in all three dimensions. This is, however, not possible for salt caverns produced with the solution method, in which case the geology and mineralogy can only be interpreted indirectly using measurements. This is similarly true for deep drill holes when complete cores are not taken (see above).

4. The long-term safety of storage and repository caverns in salt domes has been the object of numerous recent studies from the point of view of the stability of caverns in salt and the long-term isolation of the contents from the biosphere (e.g., LANGER et al. 1984; SCHOBER & SROKA 1987; LUX 1988). Two factors are of fundamental significance here (LANGER et al. 1984): Firstly, can the leaktightness of the cavern for long periods of time be proven? Secondly, does the natural pressurization in a closed cavern lead to the formation of cracks (fracturing) in the surrounding rock? Previous studies have shown that the creep of salt counteracts fracturing in the salt rock. Failure of the rock (the formation of pathways for fluids) due to cracking in the roof of a cavern can often be ruled out. LEITZKE & SROKA (1987) describe a method for the indirect monitoring of cavern convergence by measuring subsidence at the surface. A MINISTERIALBLATT (1989) with the title »Wasserwirtschaftli-

che Anforderungen an Gesteinskavernen zum Lagern wassergefährdender Stoffe«, reports on the requirements of the water resources management regarding the disposal of hazardous substances in caverns.

In spite of the previous studies on the stability of caverns dissolved in salt there are no long-term safety analyses comparable to those done for planning the disposal of radioactive waste in mined repositories. Geological factors must be considered for an optimum evaluation of the long-term safety of underground repositories for toxic substances: it must be checked whether mining techniques (mines or mined caverns) must be given preference over other techniques under complicated geological conditions (e.g., salt domes with intercalations of anhydrite or potash salt in rock salt).

## Mines

Man has been mining metals, raw materials, and gemstones for thousands of years, which resulted in voids in bodies of rocks. Hazardous anthropogenic wastes can be disposed of in these underground voids, i.e., as final repositorities (in which wastes are unretrievable).

The following distinctions are made between mined repositories for hazardous wastes:

1. mined caverns (see previous section),
2. inactive raw-material mines (e.g., the Herfa-Neurode underground repository for nonradioactive hazardous wastes), and
3. mined repositories which were planned and designed specifically for the ultimate disposal of anthropogenic wastes (e.g., Gorleben for radioactive wastes).

When using mines for the disposal of hazardous wastes, a sharp distinction must be made between former raw-material mines (case 2) and specially planned mined repositories (case 3).

The geological stratification must inevitably be taken into consideration when mining raw materials. Priority is given to nearly complete extraction of deposit and at the same time maintaining the necessary safety (HERRMANN 1983a: 123). Protection against cave-in and sheeting in a mine is only guaranteed while raw materials are being mined and for a short time thereafter. After the raw materials have been extracted, these sections of a mine are closed off and generally never entered again. Pillars for protecting the mine are dimensioned so that a maximum of raw material can be extracted (KNISSEL & FORNEFELD 1988).

For the final disposal of anthropogenic wastes, space must be created in a body of rock without detracting from the natural barrier effect of the total geological environment. Former raw-material mines cannot always meet this requirement. Frequently, the long-term stability of the rock in which such inactive mines are located cannot be assured. For example, there could be geological and compositional inhomogeneities in the rock in close vicinity to the rooms of the mine. In these cases, the suitability of such inactive mines for the disposal of anthropogenic wastes must be questioned.

However, such decisions always have to be made based on a site-specific assessment of the long-term safety (e.g., KNISSEL & FORNEFELD 1988).

*When constructing underground repositories, long-term safety is always in the limelight. In contrast, when raw materials are mined, the aim is the extraction of the greatest possible quantities of usable material, with safety being a relatively short-term factor (KNISSEL & FORNEFELD 1988).*

The differences between the three types of mines mentioned above clearly show that the possibilities for using inactive raw-material mines as hazardous-waste repositories are very restricted. In such cases, however, waste repositories could still be established in new parts of such mines. This technique appears important and promising for planning underground repositories in the future, in addition to specially constructed mines for repositories.

According to KNISSEL & FORNEFELD (1988) - supplemented by HERRMANN (see 1983a: 123) - there are three phases involved in establishing mined repositories:

1. Construction phase. The construction of individual storage chambers can take up to two years depending on chamber size and technique used (the individual chambers are meant here, not the entire mine). The construction of a complete underground system including shafts, drifts, and operating equipment can take up to ten years and even longer (e.g., Gorleben).

2. Operating phase. During this phase wastes are loaded in the individual chambers. This can take up to a year or more depending on the type and quantity of waste and chamber size. Again, this loading time refers to the individual chambers, not the entire mine. Many decades are needed to load wastes into a mined repository. An estimated 50 years will be required for Gorleben.

3. Postoperating phase. This phase is divided into two subphases:
   I. The first subphase involves the time between the loading of the chambers and the backfilling of the shafts. The time necessary for filling the individual chambers will differ depending on whether they are backfilled at the beginning or at the end of the storing procedure. In any case, this will take decades.
   II. The second subphase involves the time after backfilling of the shafts as well as the time prognosticated for long-term safety. KNISSEL & FORNEFELD (1988) define postoperating phase IIb as the time from the backfilling of the shafts to restoring the original stress state in the rock mass.

In addition to the aforementioned geoscientific aspects regarding long-term safety, there are also technical advantages to constructing repositories in previously unmined rock or in new parts of an existing raw-material mine. For example, the storage capacity of the repository can be adapted to the quantity of waste to be disposed of. The mining work on repository chambers and the loading work are to be separated in terms of space and ventilation for reasons of safety. This is easily done when establishing a new repository in an existing mine (KNISSEL & FORNEFELD 1988).

Mined repositories should be located at depths of 200-1000 m below the surface depending on the condition of the rock and site geology. Drill holes can easily be sunk to greater depths from an existing mine. Hazardous wastes, such as highly radioactive wastes with high heat emission, can be disposed of in such drill holes.

A discussion on the technical details concerning mined repositories for anthropogenic wastes is beyond the scope of this work. However, one geoscientific aspect should be pointed out again: the conditions for the long-term disposal of hazardous wastes in mined repositories (including former raw-material mines) are more favorable than those of caverns dissolved in salt. A study dealing with the possibilities for disposing of hazardous wastes in former raw-material mines should be conducted for Germany. Data on the construction of repositories extending out from existing mines would have to be emphasized in such a study. It also needs to be checked whether mined repositories specifically for nonradioactive hazardous wastes can be constructed, in addition to those for radioactive wastes (the Gorleben prototype).

Former raw-material mines are being used increasingly as underground repositories in Germany (Chapter 10). The criteria used for assessing long-term safety of such repositories will obviously differ, as in the case of repositories for radioactive wastes. Therefore, some of the repositories in former mines may very well become problems that future generations will have to deal with. Since underground repositories cannot be reclaimed, the most careful thought and planning are necessary for disposing of hazardous wastes in inactive mines (waste-disposal mining or mining and waste management).

FÜRER (1989) evaluates mined repositories based on their position with respect to groundwater. Salt mines and caverns are judged favorably in this case.

# 9 Host rocks for underground repository sites

In Germany salt rock has been and still is the preferred object for the study of the final deposition for anthropogenic wastes. In the meantime, however, other rocks are being considered and used as hosts for subsurface waste repositories (Tab. 3, see also Fig. 11). Various igneous, metamorphic, and sedimentary rock types (e.g., granites, basalts, crystalline rocks, tuffs, shales, and evaporites) are being investigated in terms of their suitability as hosts for underground repositories in other countries as well. Rock salt is the most common evaporite considered; anhydrite is seldom considered due to its total unsuitability (Tab. 3). In this context it is to be noted that previous studies of these rocks were directed primarily at the underground disposal of radioactive wastes. Studies of geological systems for the underground disposal of nonradioactive wastes have only been conducted in Germany.

**Tab. 3** Rock types studied in various countries with respect to their suitability for underground disposal of radioactive wastes. Parentheses indicate secondary importance to the respective country (from RÖTHEMEYER 1981: 779; RÖTHEMEYER & CLOSS 1981: 170; CHAPMAN et al. 1987: 232ff; NAGRA aktuell 1987; KÜHN 1988; latest developments also considered).

| Country | Rock |
|---------|------|
| Argentina | granite |
| Belgium | argillaceous rocks |
| Canada | crystalline rocks (rock salt) |
| Denmark | rock salt |
| England | granite (argillaceous rocks, evaporites) |
| Finland | granite |
| France | granite, argillaceous rocks, slates, evaporites |
| Germany | rock salt, sedimentary iron ore (for both radioactive and nonradioactive wastes), hard coal mines (nonradioactive wates) |
| India | granite |
| Italy | argillaceous rock |
| Japan | granite (shales, tuffs) |
| Netherlands | rock salt |
| Poland | rock salt |
| Spain | rock salt |
| Sweden | crystalline rocks |
| Switzerland | crystalline rocks (argillaceous rocks) |
| USA | tuffs, granite, basalt, shale, rock salt |
| former USSR | granite, rock salt, (possibly other rocks) |

All rocks of the earth take part in natural geochemical cycles. Hence, there is no »ideal« or »best« host rock for underground repositories, which was erroneously maintained during the 1970s and 1980s in Germany, and is still propagated by some in total contradiction to the scientific facts. Every rock type has certain advantages and disadvantages with respect to its geological occurrence, mineralogical and chemical composition and physical and chemical properties. The selection of a rock type to host an underground repository depends above all upon the geology of the country wanting and having to dispose of its waste outside the biosphere. In addition, social factors frequently play a role which is not to be underestimated.

The United States of America is an interesting example of the changing opinions regarding the selection of suitable rocks for the ultimate disposal of radioactive wastes. In the mid-1950s a committee of geologists and geophysicists at the National Academy of Sciences (Washington, D.C.) dealt with the possibilities for radioactive waste disposal in the USA. The committee made a statement to the effect that radioactive waste repositories in flat-lying salt beds or in salt domes could be a solution to the disposal problem (THE DISPOSAL OF RADIOACTIVE WASTE ON LAND, 1957). The salt concept was studied by the Atomic Energy Commission (AEC) and later by the Energy Research and Development Administration (ERDA) and Department of Energy (DOE). In the USA rock salt remained in the limelight as a potential host rock for the underground disposal of radioactive wastes up into the 1980s. Tuffs and their suitability as hosts for radioactive substances are now the preferred object of study in the USA. Tuffs have, among other properties, the ability to hinder or limit the dispersal of mobilized radionuclides around a repository through ion exchange.

Other rocks besides salt are also being considered in several other countries. Recent studies in the inactive Konrad iron ore mine have yielded promising results on other possibilities for establishing underground geological repositories with effective isolation of contaminants from the biosphere.

The diversity of international activities for testing suitable host rocks for the ultimate disposal of hazardous wastes will do more justice to the study of geology and the properties of natural host rocks than the reliance on one type of host. The following fact is essential to the underground disposal of anthropogenic wastes in Germany: Evaporites (rock salt) can serve as host rocks where geologically permissible, which of course does not rule out other rocks.

Scientific and technical knowledge on the underground disposal of radioactive wastes must be considered fully when planning and constructing underground repositories for nonradioactive hazardous wastes. Tab. 4 shows in which countries, in which rocks, and at which sites research on the disposal of anthropogenic wastes has been carried out or is planned. In addition, it is also indicated where radioactive and nonradioactive wastes are deposited or will be deposited.

Further activities involving the construction and use of underground repositories for nonradioactive wastes in Holland, India, and the USA (Kansas, Louisiana, Minnesota, Missouri, New York, Ohio, Texas) are described, for example, in STONE (1987).

**Tab. 4** Research and construction of underground repositories in various host rocks (geological systems) and countries. Compiled after KÜHN (1988) and supplemented after STONE (1987) and HERRMANN. [1] proposed site; [2] intended for low-level radioactive wastes; n.d., final decision on site has not yet been made; (r) for radioactive wastes; (nr) for nonradioactive wastes.

| Rock | Country | Locality, test site | Studies at tentative repository sites | Wastes were/ are being disposed of |
|---|---|---|---|---|
| **1. magmatic rocks** | | | | |
| granite | England | n.d. (r) | - | - |
| | Finland | Olkiluoto (r) | - | - |
| | France | Fanay (r) | Deux-Sèvres[1] (r) | - |
| | | Augères Mine (r) | - | - |
| | Japan | Kasama (r) | - | - |
| | Canada | Lac du Bounet (r) | - | - |
| | Sweden | Stripa (r) | Forsmark (r) | Forsmark[2] |
| | Switzerland | Grimsel (r) | - | - |
| | Spain | n.d. (r) | - | - |
| | USA | Climax Mine (r) | - | - |
| gabbro | Canada | n.d. (r) | - | - |
| | Sweden | n.d. (r) | - | - |
| diabas | Japan | n.d. (r) | - | - |
| basalt | USA | Hanford (r) | - | - |
| **2. crystalline shales** | France | n.d. (r) | Maine-et-Loire[1] | - |
| **3. tuffs** | Japan | n.d. (r) | - | - |
| | USA | Nevade Test Site (r) | Yucca Mountain[1] (r) | - |
| **4. shale** | Japan | n.d. (r) | - | - |
| | USA | n.d. (r) | - | - |
| | Spain | n.d. (r) | - | - |
| **5. argillaceous rocks** | Belgium | Mol (r) | Mol (r) | - |
| | England | n.d. (r) | - | - |
| | France | n.d. (r) | Aisne[1] (r) | - |
| | Italy | n.d. (r) | - | - |
| | Switzerland | n.d. (r) | - | - |
| **6. iron ore** | Germany | Konrad near Salzgitter (r) | Konrad near Salzgitter (r) | - |
| | | - | - | Wohlverwahrt-Nammen Mine (nr) |

**Tab. 4** continued

| Rock | Country | Locality, test site | Studies at tentative repository sites | Wastes were/ are being disposed of |
|---|---|---|---|---|
| 7. nonferrous ore | Germany | - | Rammelsberg (nr) | - |
| | | - | Bad Grund (nr) | - |
| | | - | Meggen (nr) | - |
| 8. evaporites (rock salt) salt domes | Germany | Asse (r) | Asse (r) | Asse (r) |
| | | - | Gorleben (r) | - |
| | | - | Morsleben• (r) | Morsleben• (r) |
| | | - | - | Thiederhall (nr) |
| | | - | Jemgum (nr) | - |
| | Denmark | - | Mors (r) | - |
| | Netherlands | n.d. (r) | - | - |
| | USSR | - | north of Caspic Sea | ? |
| bedded salt | Germany | - | - | Herfa Neurode Mine (nr) |
| | | - | - | Heilbronn (nr) |
| | France | n.d. (r) | Ain[1] (r) | - |
| | Spain | n.d. (r) | - | - |
| | Poland | Baltic Sea coast (r) | - | - |
| | USA | Carlsbad (r) | Carlsbad (r) | - |
| | Canada | - | - | Saskatchewan (nr) |
| | | - | - | Sarnia District, Ontario (nr) |
| evaporites (anhydrite, gipsum) | Germany | - | Obrigheim (nr) | - |
| | Switzerland | Felsenau (r) | - | - |
| evaporites (limestone, dolomite) | Germany | - | Auersmacher (nr) | - |
| | | - | Gersheim (nr) | - |
| | | - | Wellen (nr) | - |

**Tab. 4** continued

| Rock | Country | Locality, test site | Studies at tentative repository sites | Wastes were/ are being disposed of |
|------|---------|---------------------|---------------------------------------|-------------------------------------|
| 9. hard coal | Germany | - | - | Zollverein Mine (nr) |
|  |  | - | - | Consolidation Mine (nr) |
|  |  | - | Walsup | - |
|  |  | - | Monopol | - |
|  |  | - | Hugo | - |
| 10. permafrost area | Canada | - | - | Yukon Territory (nr) |

# 10  Underground repositories in Germany

In Germany underground repositories for the ultimate disposal of anthropogenic wastes are in use and being planned. Only nonradioactive wastes are currently being disposed of. The Asse II salt mine near Wolfenbüttel (Niedersachsen), where low- and medium-level radioactive waste was disposed of between 1967 and 1978, was the first repository used for such wastes. The underground repository Morsleben in the former GDR was and is also used for the disposal of low level radioactive wastes.

Inactive and active raw-material mines have been the primary object of use and study for underground repositories. Inactive raw-material surface mines are also used for disposing of power plant wastes and household trash. The only sites planned and studied for the exclusive purpose of waste disposal are the Gorleben salt dome (for radioactive wastes) and the Jemgum salt dome (caverns dissolved in rock salt for nonradioactive wastes; see Fig. 11). It is obvious why raw-material mines are attracting more and more interest as possible sites for nonradioactive waste disposal whereby the existing storage space for the wastes and the associated cost-savings are obviously a decisive factor. However, these advantages are only relevant to the present. Geoscientific criteria concerning the long-term safety of inactive raw-material mines as repositories are much more important. In this respect, not all the space available in various mines is equally suitable for the long-term isolation of hazardous wastes from the biosphere. This fact must be emphasized repeatedly. The previous and current studies on long-term safety conducted at Gorleben and the Konrad mine before actual radioactive waste disposal are exemplary. They should serve as a measure for evaluating the long-term safety of underground repositories for nonradioactive wastes as well. The toxic effect of nonradioactive wastes lasts much longer than that of radioactive material (see Chapter 5). Unfortunately, practically no comprehensive study and prognoses especially regarding this critical aspect have been considered in evaluations of the safe disposal of nonradioactive wastes in raw-material mines. Hence, underground disposal of nonradioactive wastes should be centrally planned and above all monitored as in the case of radioactive wastes.

In Germany the underground repositories for anthropogenic wastes are already being planned and constructed in the following rocks and raw-material mines: evaporites, hard coal, ferrous and nonferrous ores, limestone, and dolomite (Tab. 5). Hazardous anthropogenic wastes were first disposed of primarily in evaporite bodies. In recent years, inactive hard-coal mines have also been used and studied to an increasing extent as repositories above all for power plant wastes (e.g., filter dust; PLATE 1988). The same is true for inactive ore mines; here a differentiation is made between mines for ferrous and nonferrous metals (Tab. 5). Even underground gypsum, limestone, and dolomite mines have been used for disposing of power plant wastes. At some underground mines in Germany the term »Endlagerung« (final disposal) for certain power plant wastes has been replaced by the colloquial German word »Ver-

36

1 Gorleben (evaporites)     2 Morsleben (evaporites)
3 Asse (evaporites)         4 Konrad (iron ore)
5 Herfa Neurode (evaporites) 6 Heilbronn (evaporites)
7 Jemgum (evaporites)       8 Zollverein (hard coal)
9 Wohlverwahrt-Nammen (iron ore)

**Fig. 11** Underground repositories in Germany which have been used, are in use, or are being studied and planned. Only a selection of the sites being studied and planned is given. Dot with open circle indicates radioactive wastes; two open circles indicate nonradioactive wastes.

satz« (backfill). It is pointed out that the materials disposed of in this way have toxic-element contents many times that of the mean composition of polluted soils (Tab. 2). For the reader not acquainted with mining, the technique of backfilling involves the refilling of an excavation or mine after removal of the desired material. This is done for reasons of safety and should, for example, minimize or prevent entirely subsequent damage to the land surface above the mine (e.g., subsidence). A wide variety of materials is used for backfilling purposes. It can involve simply replacing the rock material removed temporarily during mining. The residues from the processing of raw materials are also used occasionally as backfill, an example being the rock salt and clay minerals left over from the processing of potash salt. It is primarily the clay minerals that are used in various mines for backfilling the underground voids, but the bulk of this material ends up on waste banks at the surface. This is only done with materials which do not contain toxic element concentrations. The chemical composition is the basis for differentiating the material common in potash and salt mining from power plant residues declared as backfill. The final disposal of residues containing high amounts of toxic elements and compounds (Tab. 2) should always be referred to as such and never as backfill, as is often done (Tab. 5). Wastes for underground disposal can only be controlled properly in connection with the necessary planning procedures for a repository.

In spite of the use of different host rocks for the underground disposal of anthropogenic wastes in Germany, evaporites will continue to play an important role in toxic waste disposal in the future. This is equally true for nonradioactive hazardous wastes and for radioactive materials of all categories. Therefore, the existing and planned repositories in evaporites of Germany will be presented in the following.

## Herfa-Neurode

The currently most important underground repository for the final disposal of nonradioactive hazardous wastes is the inactive part of the Herfa-Neurode mining field of the Wintershall potash mine in Heringen (Werra-Fulda mining district). Disposal work proceeds along side that of the potash works, but is otherwise totally separate from mining activities. The horizon in which the wastes are being placed is at a depth of 700 m, i.e., at the level of the currently mined Thüringen potash salt seam (Fig. 12). The rocks of the Thüringen potash seam are the oldest K-Mg mineral associations which formed about 250 million years ago in the Werra sequence of the central and western European Zechstein evaporites. The wastes are being placed in the former rooms which were prepared for this purpose. After the rooms are filled with waste drums, the access openings are walled off.

The shallow-dipping evaporite sequence is about 300 m thick and is overlain by about 400 m of Buntsandstein. Interbedded with these rocks is a collective 60 m of clay-mineral-bearing beds, which are undisturbed and act as a seal against water and other solutions between the repository and the biosphere (Fig. 12; JOHNSSON 1983, 1985). The underground repository at Herfa-Neurode is a good example of a multibarrier geological system.

Since the repository began operation in 1972, about 1 000 000 t of hazardous waste has been disposed of here. The annual emplacement capacity at Herfa-Neurode is presently limited to about 160 000 t due to the availability of the Herfa shaft. However, the available underground space would allow an even larger quantity of wastes per year. According to the calculations of BIELER & CLAUS (1988) 5-8 million t of hazardous waste could be disposed of annually in the salt mines of the Werra-Fulda mining district. It must be mentioned, however, that this figure does not take into account any information on the actual suitability of the sites and underground space

**Tab. 5** Underground repositories planned and in operation for radioactive and nonradioactive wastes in Germany. Compiled by HENNES (1989) and supplemented by HERRMANN. m, mine; c, cavern; nr, nonradioactive; r, radioactive; ni, no information. * indicates quantities already disposed of.

| No. | Rock | Mine cavern | Locality | Federal state | Kind | Deposition | Wastes | Situation | Capacity [m³] |
|---|---|---|---|---|---|---|---|---|---|
| 1 | evaporites | mine | Herfa-Neurode | Hessen | deposit | nr | hazardous industrial waste | working | 160 000/year |
| 2 | (rock salt) | mine | Heilbronn | Baden-Württemberg | deposit | nr | incineration | working | 100 000/year |
| 3 | | cavern | Jemgum | Niedersachsen | deposit | nr | hazardous industrial waste | planned | Σ 2 800 000 |
| 4 | | mine | Thiederhall | Niedersachsen | deposit | nr | sludges | deposit | Σ 380 000* |
| 5 | | mine | Asse | Niedersachsen | deposit | r | radioactive wastes | deposit | Σ 25 000* |
| 6 | | mine | Gorleben | Niedersachsen | deposit | r | radioactive wastes | exploration | Σ 5 000 000 |
| 7 | | mine | Morsleben | Sachsen-Anhalt | deposit | r | radioactive wastes | deposit | Σ 14 300 * |
| 8 | gypsum | mine | Obrigheim | Baden-Württemberg | deposit | nr | power plant wastes | working | ni |
| 9 | limestone, | mine | Auersmacher | Saarland | backfill | nr | power plant wastes | application | Σ 4 000 000 |
| 10 | dolomite | mine | Gersheim | Saarland | backfill | nr | power plant wastes | working | Σ 7 000 000 |
| 11 | | mine | Wellen | Rheinland-Pfalz | backfill | nr | power plant wastes | application | ni |
| 12 | hard coal | mine | Zollverein | Nordrhein-Westfalen | deposit | nr | power plant wastes | working | Σ 155 000 |
| 13 | | mine | Consolidation | Nordrhein-Westfalen | backfill | nr | power plant wastes | working | ni |
| 14 | | mine | Walsum | Nordrhein-Westfalen | backfill | nr | power plant wastes | investigation | ni |
| 15 | | mine | Monopol | Nordrhein-Westfalen | backfill | nr | power plant wastes | investigation | ni |
| 16 | | mine | Hugo | Nordrhein-Westfalen | backfill | nr | power plant wastes | investigation | ni |
| 17 | iron ore | mine | Wohlverwahrt-Nammen | Nordrhein-Westfalen | backfill | nr | power plant wastes | working | 100 000/year |
| 18 | | mine | Konrad | Niedersachsen | deposit | r | radioactive wastes | exploration | Σ 1 000 000 |
| 19 | nonferrous ore | mine | Rammelsberg | Niedersachsen | deposit | nr | processing wastes | investigation | 100 000/year |
| 20 | | mine | Bad Grund | Niedersachsen | backfill | nr | power plant wastes | investigation | ni |
| 21 | | mine | Meggen | Nordrhein-Westfalen | backfill | nr | power plant wastes | application | ni |

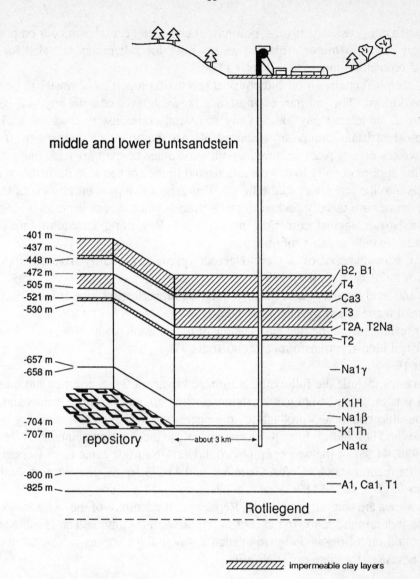

**middle and lower Buntsandstein**

| | |
|---|---|
| -401 m | |
| -437 m | |
| -448 m | |
| -472 m | B2, B1 |
| -505 m | T4 |
| -521 m | Ca3 |
| -530 m | T3 |
| | T2A, T2Na |
| | T2 |
| -657 m | Na1γ |
| -658 m | |
| | K1H |
| -704 m | Na1β |
| -707 m | K1Th |
| repository  ◄— about 3 km —► | Na1α |
| -800 m | |
| -825 m | A1, Ca1, T1 |

**Rotliegend**

▨▨▨▨▨ impermeable clay layers

**Fig. 12** Profile of the rock sequence at the Herfa-Neurode repository (FINKENWIRTH & JOHNSSON 1980).

with respect to the long-term safe isolation of the wastes from the biosphere (see Chapter 11). Furthermore, such figures do not give any indication as to whether the mines are technically capable of handling such large quantities of waste (conveying capacity of the shafts, underground transport system, separation of areas being mined from repository areas). Hence, there is the danger that hopes could be raised which could not possibly be fulfilled due to geological, technical, and social reasons, especially when nongeoscientists (e.g., politicians) cite such general estimates on potential repository space in mines. This is true not only for salt mines but also for other potential repository mines (see Chapter 11).

The chemical composition and physical and toxic properties of wastes to be stored must be known. The substances must not be explosive, contain any self-igniting compounds, or release any gases. Only solid and nonradioactive wastes packed in mechanical resistant drums are disposed of. An important requirement is that the stored wastes do not react or interact with each other or the host rock (mainly rock salt). This applies equally to dry conditions and to the conceivable intrusion of saline solutions into the repository in the future. This aspect is important in view of the fact that the wastes are usually packed in metal drums which do not serve as a long-term effective barrier against corrosion in evaporites. Regarding long-term safety such containers are only suitable for transport.

The wastes disposed of at Herfa-Neurode originate primarily from the following sectors of industry (e.g., KIND 1991):
1. iron and steel industry: heat-treating, galvanic metallization, smelteries 30%,
2. residual wastes from the chemical industry 20%,
3. filter residues from the flue gas cleaning of incineration plants 40%,
4. electrical industry: transformers, condensers 9%,
5. other 1%.

These wastes include the following: chlorinated hydrocarbons; arsenic- and mercury-bearing wastes; wastes from galvanization, distillation, and filtration; paints and dyes; tars; unusable chemicals; products of the pharmaceutical industry; cleaning agents; pesticides and herbicides; filter residues from refuse incineration plants.

In 1989, 41.5% of the wastes emplaced in Herfa-Neurode came from Hessen, 47% from other federal states of West Germany, and 11.5% from other Western European countries (e.g., KIND 1991).

The wastes are carefully registered. Representative samples of the hazardous wastes are kept underground, allowing checks at any time in the future. The Herfa-Neurode underground repository is designed in such a way that the wastes can be taken out of store if new uses for these materials arise.

## Heilbronn

Since 1987 nonradioactive wastes have also been disposed of in a second salt mine, namely, in the Heilbronn rock salt mine of the Südwestdeutsche Salzbergwerke AG. These evaporite deposits were formed during the Triassic around 200 million years ago. Stratigraphically this bedded rock salt belongs to the Middle Muschelkalk, and in the area of Heilbronn the evaporite sequences are about 40 m thick and lay at a depth of about 200 m. The units of the Middle Muschelkalk are made up of the following strata (according to LOTZE 1938: 577; see also Fig. 13):

Upper Muschelkalk

| | | | |
|---|---|---:|---|
| | upper dolomitic beds: dolomite, limestones, oolite | 14.4 | m |
| | upper anhydrite beds: anhydrite, argillaceous layers | 31.8 | m |
| Middle | upper argillaceous beds: clays with anhydrite | 9.2 | m |
| Muschelkalk | bedded rock salt with 0.6 m of intercalated anhydrite | 40.4 | m |
| | lower anhydrite beds: anhydrite, clays | 2.2 | m |
| | lower dolomitic beds: dolomite | 4.5 | m |

Lower Muschelkalk (Wellenkalk)

Rock salt has been mined here for 100 years. As it appears now, rock salt will continue to be mined here for another 100 years. According to data of the state mining authorities there is space for 25 million m³ of waste in this mine (UNTERIRDISCHE MÜLLDEPONIEN 1988). It is assumed that 80 000 to 100 000 t of wastes can be disposed of annually.

The underground repository is situated in rooms of the mine exploited about 20 years ago. Repository and rock salt mining operations are separated spatially and organizationally and have different ventilation systems.

The filter residues from the flue gas cleaning of the Göppingen refuse heating plant are disposed of here. In the future, filter residues from flue gas cleaning from other refuse incineration plants in Baden-Württemberg are also to be emplaced in Heilbronn.

The filter residues of flue gas cleaning are filled into metal drums with plastic liners (BURGBACHER et al. 1986). As at Herfa-Neurode, such containers are only suitable for transport and are not effective barriers with regard to the long-term safety of the repository for hazardous nonradioactive wastes in the Heilbronn rock salt mine.

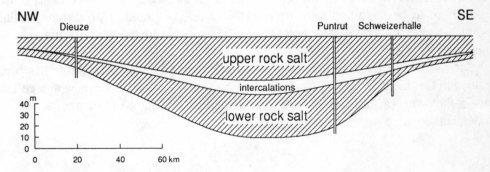

**Fig. 13** Schematic east-west cross section through the salt basin of the Middle Muschelkalk in southern Germany (from LOTZE 1938: 582).

## Kochendorf, Stetten

In Baden-Württemberg the salt mines in Kochendorf and Stetten are also being considered for disposal of nonradioactive wastes, in addition to the Heilbronn repository. Collectively, there would then be approx. 45 million m³ of underground space available for waste disposal (UNTERIRDISCHE MÜLLDEPONIEN, 1988).

## Obrigheim

Calcium sulfate - in the form of gypsum ($CaSO_4 \cdot 2\ H_2O$) or anhydrite ($CaSO_4$) - is also a constituent of the rocks in a marine evaporite sequence. In Baden-Württemberg the Heidelberger Zement AG is planning to use the gypsum mine in Obrigheim as a final repository for the ash of the flue gas desulfurization plant of the Heilbronn and Karlsruhe hard-coal-fired power plants. Initial studies have already been conducted (UNTERIRDISCHE MÜLLDEPONIEN 1988; Tab. 5).

## Jemgum

Along the north sea coastal area in Niedersachsen plans are being made for solution caverns in a salt dome for disposal of certain nonradioactive solid wastes. The Bunde, Jemgum, and Etzel salt domes were first discussed as possible sites for repository caverns (Fig. 14). The Jemgum salt dome was selected for the pilot project. Here, 20 caverns with a collective volume of 2.8 million m³ are to be created by solution mining for disposing of a few selected types of waste (see also Chapter 8). The individual caverns should have a usable volume up to 200 000 m³ with a diameter of 50 m and a height of about 200 m. The maximum depth for the cavern roots should be about 800 to 1000 m (e.g., SCHAAR 1989). Salt substances with salt character as well as fly ash and similar matters are under consideration as wastes. The material is to be emplaced in the caverns without being conditioned (see Fig. 10). The contents of the cavern are to be solidified into a monolithic block by addition of a solidification agent. Chapter 8 contains aspects that are also to be considered when filling caverns with wastes.

Sufficient experience in the construction of caverns by solution mining has been gained in Germany. In the 1970s, for example, 30 underground caverns with a collective volume of 12 million m³ were constructed in the Etzel salt dome for storing federal crude oil reserves.

## Asse

Asse is a salt dome located about 10 km southeast of the city of Wolfenbüttel (Niedersachsen) in the northern foreland of the Harz Moutains. The salt rocks are overlain by gypsum and hydrous gypsiferous clays up to 150 m deep.

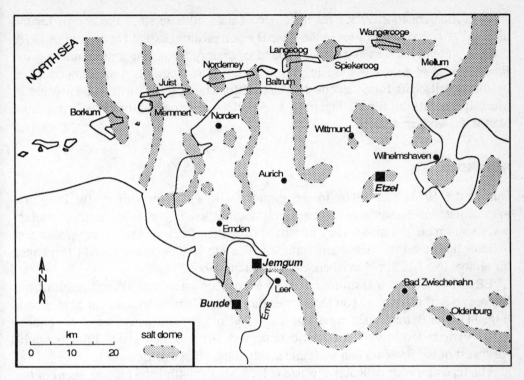

**Fig. 14** Salt domes in the northern part of Niedersachsen considered for the construction of underground repositories for nonradioactive wastes.

Work on shaft I began in 1899, already reaching its final depth of 375 m in 1900. The mine was abandoned in 1906 due to invading water and solutions.

The sinking of shaft II (Asse II) began in 1906. It had a depth of 750 m. Salt ceased to be mined here not until 1964. There was obviously no excessive influx of water here from the time mining operations began until they were concluded. After the mine was shutdown, nonradioactive and low- and intermediate-level radioactive wastes with low heat generation were placed in Asse II between 1967 and 1978.

The Asse III shaft was sunk beginning in 1911 and had a depth of 730 m. After solutions invaded the 725 m level, the lower portion of the shaft was sealed off at the 680 m level with a concrete plug. The bottom of the shaft remained dry thereafter until its shutdown (BAUMERT 1928: 32-33).

Approximately 25 000 m³ of radioactive wastes has been disposed of in Asse II. It must always be emphasized that the wastes emplaced in the Asse II mine, which is officially designated as a test repository, are not retrievable. This means that in the case of a conceivable accident the radioactive wastes could not be removed completely.

No further radioactive wastes have been emplaced here since the permit for the disposal of radioactive wastes in the Asse II test repository expired at the end of 1978. Asse II presently serves as an underground laboratory for testing and monitoring the final disposal of radioactive wastes in salt domes. The facility is now being operated by the Gesellschaft für Strahlen- und Umweltforschung (GSF) with headquarters in Munich. The Institut für Tieflagerung, Braunschweig, as part of the GSF, is responsible for the work in Asse II.

## Morsleben

During the 1970s and 1980s in the former GDR a deep repository for low- and intermediate-level radioactive wastes was established in the inactive Bartensleben potash and rock salt mine. They consisted of wastes from nuclear power plants and radionuclides used in science and industry. The site is abbreviated ERAM (Endlager für radioaktive Abfälle Morsleben, e.g. HERRMANN 1992).

ERAM is located 600 m northwest of Morsleben and about 1.5 km north of the Hannover-Berlin Autobahn on the former East-West German border near Marienborn (Fig. 11). The Bartensleben mine with the repository for radioactive waste is connected at various levels with the Marie mine and shaft located about 1.6 km north-northwest of the Bartensleben shaft and south of the village of Beendorf.

The repository for radioactive waste is located at the 4th level (506 m deep) of the Bartensleben mine. From March 1978 to December 1990 around 14 300 m³ of low-level radioactive waste and 6892 radioactive sources have been deposited there (BfS Infoblatt Feb. 1991). Following the reunification of Germany on 3 October 1990 the Bundesamt für Strahlenschutz (BfS) assumed responsibility for the operation of ERAM. A new geotechnical concept is being worked out to assure the long-term safety of ERAM.

In the Marie mine, connected with Bartensleben, 5757 t of wastes from the galvanic industry have been stored at 360 m depth. They are to be removed and brought to another location.

## Gorleben

In contrast to Asse II and Morsleben, a special repository mine for radioactive wastes of all categories is being tested in the Gorleben salt dome in an evaporite body which has not yet been mined. From a geoscientific point of view, this is a considerably safer method than disposing of heat-generating radioactive wastes and of radionuclides with long half-lives in old mines, originating from the early days of mining.

If the Gorleben salt dome should prove to be suitable as an underground repository, the prototype of a special repository mine would be established here. A drift is to be driven at 830 m depth with entrances to repository chambers for heat-generating and non-heat-generating radioactive wastes. From the 830 m level, underground drill holes are also to be sunk for storing containers with high heat-generating radioactive wastes (see Fig. 15). The repository rooms are to be located between 850 and 1100 m depth.

In the rock salt beds of the planned Gorleben repository 5 million m³ of space can be created for disposing of radioactive wastes of all categories from nuclear energy, nuclear technologies, research centers, state collection sites and industry. The disposal is calculated to take 50 years. When disposal is finished, a complete filling of the underground rooms and shafts is intended. This is very important to the long-term safety of the repository.

The discussion on the suitability of the Gorleben salt dome for the construction of a repository for radioactive wastes can be summarized as follows:

Previous studies indicate that there are saturated solutions in the anhydrite and rock salt beds down to depths of 2000 m and perhaps even deeper around the flanks of the salt dome. The rocks overlying the salt dome and the evaporite body itself were highly strained during the Elster glaciation that today solutions and Quaternary sediments directly overlie the evaporites in the central part of the locality. Consequently, the barrier system of the Gorleben salt dome cannot be compared directly with the flat-lying evaporites at Herfa-Neurode. In contrast to Herfa-Neurode, the easily water soluble rock salt of the dome must act as the essential barrier in Gorleben due to the lack of a continuous argillaceous rock layer between the salt dome and the biosphere. There are several hundred meters of rock salt between the planned underground repository chambers and the outer limit of salt dome above and on the flanks. The question is whether or not the barrier effect of the rock salt will always be maintained under the influence of the great heat generation of the high-level radioactive wastes.

Various studies have been conducted on the long-term safety of an underground repository in the Gorleben salt dome. Among other aspects, the processes which occurred in the geological past and which have altered the original composition of the salt dome locally and over limited periods of time are considered (see Section III).

Shafts are currently being sunk in Gorleben since most surface exploration has been completed.

## Thiederhall

The former Thiederhall potash mine in Salzgitter-Thiede began operation in the Thiede salt dome at the end of the 19th century (shaft I (1), 1885-1891) and the beginning of the 20th century (shaft II (1a), 1900-1905; shaft III (2), 1912-1916) following initial exploratory drilling as early as 1872. The shafts are connected at various levels. The deepest drift was driven at a depth of 600 m from shaft III (2). The mined potash salt was a carnallitic rock with about 9% $K_2O$ from the Staßfurt potash seam. Due to the economic situation, operation at the Thiederhall mine was ceased in 1924 (e.g., HOF-RICHTER 1980b).

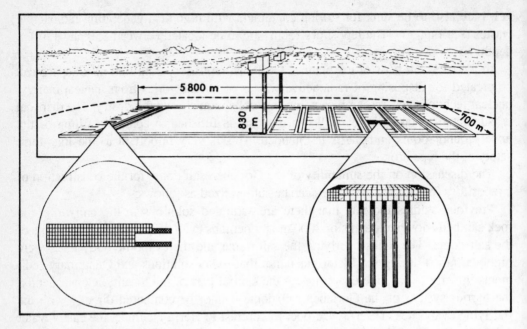

**Fig. 15** Schematic representation of the planned underground repository in the Gorleben salt dome. From MEMMERT (1983).

Between 1977 and 1984 the Volkswagen factories in Braunschweig and Hannover deposited approx. 380 000 m³ of industrial sludge into underground rooms of the mine. The Thiederhall underground repository was shut down in 1987. These sludges consisted of up to 90 - 99% water. The solids were primarily mixed enamel sludges, which were in suspension due to their fineness and did not settle. A collective volume of 800 000 m³ was supposedly available for the liquid wastes in the former room of the mine.

The sludges were fed into the mine though a specially lined drill hole at a depth of about 283 m, which extended down to about 600 m depth. The shafts were sealed off before the sludges were filled into the rooms of the mine.

## Other tentative underground repositories in evaporites

It was already pointed out in the above discussion on the underground repository in the Heilbronn salt mine that in the future even more raw-material mines will surely be investigated regarding their suitability as repositories for hazardous wastes. Basically inactive mines should be checked with respect to their suitability as an underground repository before their final closing, filling, or flooding (e.g., Hope salt mine).

Repositories for the sole purpose of waste disposal in previously unmined salt domes are also possible. The evaporites in Niedersachsen (and Schleswig-Holstein?) are being considered for this.

In terms of long-term safety, the construction of specially designed repository mines allows the technical measures and geology to be taken into account from the very beginning. This is, in contrast, not possible in the case of raw-material mines. Specially designed repository mines also have advantages over salt caverns produced by solution mining as regards the long-term safe disposal of hazardous wastes in salt mines.

## Konrad

To supplement the underground repositories in evaporites discussed above, the disposal of radioactive wastes in the former Konrad iron ore mine will be presented. Non-heat- and low-heat-generating radioactive wastes are to be disposed of in the Konrad iron ore mine which has not been in use since 1976. The mine is located near Salzgitter-Bleckenstedt. The Konrad 1 (1232.5 m deep) and Konrad 2 (997.5 m deep) shafts were sunk between 1958 and 1962. The iron ore mined here is part of a sedimentary-marine iron-ore deposit extending through the area of Peine, Salzgitter, and Salzgitter-Gifhorn. There are two iron ore horizons each up to 13 m thick at a depth of 850 - 1100 m. Predominantly shales, marls, and limestones underlie and overlie the iron ores. The Konrad mine remained dry during operation. The influx of water as the shafts were being sunk was stopped by grouting. Geological and mine-engineering studies on the safety of the Konrad mine, as a possible repository for radioactive wastes, have been conducted since 1975. The isolated solutions found in the Konrad mine today are formation waters and residual backfill solutions (for summary see HERRMANN 1983a: 167 ff).

The mine has a collective volume of 2.5 million m³, and a usable volume of 1 000 000 m³. The calculated disposal rate is 20 000 to 40 000 m³ per year over a period of 20 years.

# 11 Limitations of underground repositories

The possibilities for the long-term safe disposal of hazardous waste underground are limited above all by the quantities of waste and the number of geologically suitable and socially acceptable sites in a densely populated industrial country. Geologically suitable sites are defined as natural systems whose geology and composition assure the long-term safe isolation from the biosphere of the wastes deposited in underground repositories.

The situation regarding the radioactive wastes to be disposed of in Germany will first be discussed. With regard to quantity, the wastes are comparatively easy to evaluate. There are presently two sites both located in Niedersachsen: the Gorleben repository mine which is still in the reconnaissance stage and the former Konrad iron ore mine which is currently being prepared for use as an underground repository. If necessary, a total volume of up to 5 million m³ could be created in the Gorleben repository. About 1 000 000 m³ of usable volume for a repository can be obtained in the Konrad mine (Tab. 5). Since only about 14 000 m³ of radioactive wastes of all categories is estimated to accumulate annually, there would be underground space for radioactive wastes for decades to come when the planned undergound repositories Konrad and Gorleben are commissioned. Even the wastes of power plants, research centers, and state collection sites (approx. 200 000 m³) presently in intermediate storage awaiting ultimate disposal do not change this situation (see Fig. 16).

Yet, the situation is very different for the quantities of nonradioactive wastes which are one hundred times as large and also have to be isolated safely from the biosphere. For the following considerations the favorable case is presumed in which the annual quantity of hazardous waste can be reduced from the current 4.9 million t (according to the recent technical instructions for hazardous waste, there are about 15 million t per year) to 2.2 million t (of which 1.5 million t are solid) by the turn of the century (see Chapter 4.2). At least a portion of this waste must be disposed of in underground repositories, but there is no consistent information on the exact quantity. Of the 15.6 million t of hazardous waste for 1984 6.6% (i.e., 1.03 million t) should have been disposed of in underground repositories in West Germany (Fig. 4). Other estimates yield 360 000 t per year (see Chapter 4.2). Considering the new German technical instructions for hazardous waste, the assumption that about one million t of solid hazardous wastes must be stored in underground repositories annually appears realistic. This figure probably presents a minimum because one million t of hazardous waste is projected alone for deposition in salt caverns (GOVERNMENT PRESS RELEASE 1989).

Of the underground repositories in operation, Herfa-Neurode, Heilbronn, and Wohlverwahrt-Nammen could accommodate about 300 000 m³ (Tab. 5, Fig. 11). Regarding this figure it is to be remembered that approx. 200 000 t of residues from refuse-incineration and power plants are to be disposed of in Heilbronn and Wohlver-

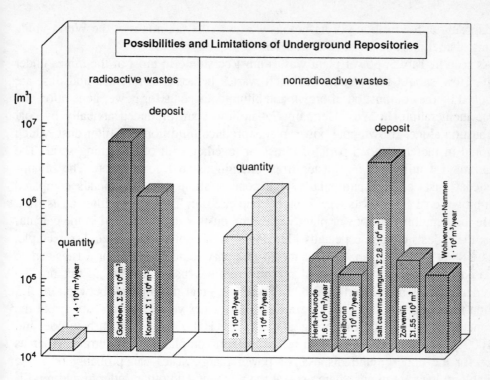

**Fig. 16** The quantities of radioactive and nonradioactive wastes accumulating annually in West Germany and the possibilities for their long-term safe disposal underground.

wahrt-Nammen. Only the Herfa-Neurode repository has been available for toxic wastes from the chemical and other industries (see Chapter 10). This means that the three aforementioned underground repositories now in operation cannot handle even half of the wastes which need to be disposed of each year.

The situation is different when the total existing volume of the three mines is considered in the calculation for repository purposes. In this case 5 - 8 million t of hazardous wastes could be disposed of in the space created annually in the active salt mines of the Werra-Fulda mining district. In the Heilbronn salt mine there is room for 25 million m³ of waste (Chapter 10), and in the Wohlverwahrt-Nammen there is 6 million m³ of available space at depths of 30 - 350 m to be used for waste disposal (HENNIES 1989). Hence, Heilbronn and Wohlverwahrt-Nammen have a collective volume of 31 million m³, plus the 5-8 million m³ or t per year of the salt mines in the Werra-Fulda mining district. An additional total volume of 22 million m³ for waste disposal presumably available in ten bituminous-coal mines of the Ruhr region (BIELER & CLAUS 1988). Again, the uncertainty of such calculations must be emphasized (see also Chapter 10, Herfa-Neurode). Independent of the geological requirements for assuring the long-term safety of such underground repositories, the technical aspects of disposing of 5-8 million t of hazardous wastes in the salt mines of the Werra-Fulda mining district must be considered in detail.

As seen in Tab. 5, power plant wastes must be stored in most of the mines under study. How should the volumes of such wastes be assessed? Ashes and dust are produced by the combustion of brown and bituminous coals for power generation and refuse incineration. In West Germany 7.4 million t are produced annually by coal combustion alone. Brown-coal power plants produce 6 million t of filter dust, which is stored in the abandoned parts of mines or together with other mining spoil. The remaining 1.4 million t are wastes from bituminous-coal combustion. The bituminous-coal dusts contain particularly high concentrations of toxic metals compared with brown-coal filter dusts (see Tab. 2, Chapter 5). In Tab. 1 only filter dusts from refuse incineration - not power plant wastes and dusts - are considered in the calculations; these dusts contain especially high amounts of contaminants (see, for example, HERRMANN et al. 1985, Tab. 2). This means that only some power plant wastes can be disposed of safely in the potential repository mines listed in Tab. 5, but not the hazardous wastes mentioned in Tab. 1. There is a great discrepancy between the 1.5 million t of solid nonradioactive hazardous wastes per year given in Tab. 1 and the current possibilities for disposing annually of 160 000 t in Herfa-Neurode and 100 000 m³ of filter dust, etc., from refuse incineration plants in Heilbronn. Whereas space for the underground disposal of power plant wastes can probably be made available in inactive mines, there are - at the moment - no other underground repository sites similar to Herfa-Neurode in site.

And what is the situation regarding salt caverns? The planning for 20 caverns with a total volume of 2.8 million m³ in the Jemgum salt dome is discussed in Chapter 10. This project is planned for special types of hazardous wastes. According to a government press release of 1989, there should be approx. 1 million t of waste suitable for final disposal in salt caverns per year. There have been no comprehensive scientific investigations on the suitability of certain types of hazardous wastes for final disposal in salt caverns. Extensive studies of the long-term safe behavior of the wastes in the salt caverns, especially regarding geoscientific and chemical aspects, still have to be conducted as well. The following example should illustrate how limited the volume is which is created in such caverns in comparison to the quantities of wastes to be disposed of underground.

If, for example, the quantity of 1 million t given in a government press release of 1989 would be disposed of annually in salt caverns, cavern facilities, like those planned for the Jemgum salt dome, could be filled every three years. However, due to problems of selection and public acceptance of such sites it is unrealistic to presume that new caverns could be constructed so quickly.

The quantities of hazardous wastes to be disposed of underground and the possibilities which actually exist are shown in Fig. 16. The filter dusts from coal-fired plants are not contained in the 0.3 - 1 million m³ of wastes to be disposed.

It is obvious that the problem of disposing of nonradioactive wastes cannot be solved with the currently available surface and underground repositories. On the other hand, it is sure that the measures necessary for reducing the amounts of wastes produced will have to far exceed those planned if the large gap between waste volume and repository space for nonradioactive wastes is to be closed.

There is about 80 million $m^3$ of surface repository space still available in West Germany (GOVERNMENT PRESS RELEASE 1989). This means that with 5 million t of hazardous wastes accumulating each year, the space in surface repositories would be exhausted in 10 - 15 years. These figures do not consider the fact that every surface repository for hazardous wastes is to be evaluated cautiously with respect to the long-term effect of the wastes and the long-term safety of the repository.

There is the danger that German mining companies might like to supplement their function as suppliers of raw materials by acting as waste repositories (e.g., JEZIERSKI 1989). The authorities and decision-makers must observe this trend carefully so that long-term safety is sacrificed for profit.

# 12 Assessment of the present situation

Because of the present imbalance between the quantities of wastes and the possibilities for their disposal, it must be concluded that the problem of disposing of nonradioactive hazardous wastes cannot be solved with either surface or underground repositories. On the other hand, one thing is sure: the necessary measures for reducing waste production will have to far exceed those planned to correct this imbalance between waste volume and available repository volume.

Wastes can be processed physically and chemically in the following ways: neutralization, precipitation, detoxification, size reduction, sorting, dewatering, filtration, gravity separation, and others. The aim of such methods is to reduce the toxic or polluting effect of the wastes and to generate reusable wastes. In addition, the thermal processing of solid and liquid wastes is also a possibility for reducing hazardous waste qualities. In the incineration process residues in the form of slag, ash, filter dust, and salts are produced which frequently contain certain toxic elements in higher concentrations and thus must be disposed of long-term safety (e.g., Tab. 2).

Worldwide experience with the underground disposal of radioactive wastes over the last two decades has shown which comprehensive and long-term studies have to be conducted before statements and decisions on the suitability of repository sites are possible. It is not always known whether the mines with millions of cubic meters of space for disposing of nonradioactive wastes are checked with the same thoroughness regarding their long-term safety as those underground repositories for radioactive wastes. Repository mines must not be allowed to become hazards to the biosphere after several decades or centuries posing problems like those at the surface today.

Even though the amounts are comparatively small, it is wrong to mention underground repositories only in connection with the disposal of radioactive wastes, as is still done today. A similar or even greater problem, particularly with regard to planning, is the safe long-term isolation of the multitude of toxic nonradioactive wastes. A great need for far-sighted planning remains to be satisfied. In this context, the work presented here should help stimulate the discussion process.

*Critical self-regulation of every individual step necessary in constructing underground repositories is a must for scientists, decision-makers, and the public interested in constructive solutions. This is imperative because we will not be able to evaluate the proper isolation of underground repositories constructed today. Only future generations will be in this position. In other words, the mistakes we make today while constructing underground repositories will confront the people of the forthcoming centuries. This aspect should never be let out of sight when planning and using underground repositories.*

# Part II

# Fluids in Marine Evaporites

# 13 Salt solutions and fluid inclusions in marine evaporites

## 13.1 Scientific fundamentals

The mining and use of marine evaporites has shown that they contain fluid phases in the form of aqueous solutions, gases, and liquid hydrocarbons (oil, condensates), in addition to solids. Salt solutions are found in both flat-lying (bedded salt) and steeply inclined (salt domes) evaporite bodies (see v. BORSTEL 1992).
Solutions are found
1. on the surfaces of fissures and cracks and in caverns primarily in anhydrite beds, but also in saliferous clay, potash salt rocks, and rock salt. The volumes of such salt solutions vary from a few ml up to several 1 000 m³; and
2. as fluid inclusions in minerals such as halite, sylvite, carnallite, polyhalite, kainite, kieserite, anhydrite, and gypsum. Polyphase inclusions (liquid-gas, liquid-solid) are also observed. The size of such fluid inclusions varies from a few to several 100 μm. The volume comprises up to 2 % of the evaporites.
Studies of recent evaporites have shown that small amounts of fluids can be trapped in the salt minerals already during crystallization in the basin of deposition. Fluid inclusions of this type are referred to as primary inclusions (e.g., ROEDDER 1984).

Even after deposition of the evaporite, unsaturated solutions from neighboring and/ or overlying rocks can penetrate into the evaporite and cause extensive mineral reactions (solution metamorphism). In addition, increased temperatures can cause saturated salt solutions to form from hydrate minerals within the evaporite bodies (thermal metamorphism). Such solutions can be fixed in solid reaction products in the form of fluid inclusions, or squeezed out and stored on the surfaces of fissures and cracks and in other voids (»Gebirgslösungen«).

Although it is easy to sample and analyze such »Gebirgslösungen«, the quantitative determination of the composition of fluid inclusions in evaporite minerals has long been problematic. A brief summary of the most important methods for the microanalysis of fluid inclusions in salt minerals is given in the following:
1. Indirect methods (without destroying the inclusions).
   1.1 Measurement of the freezing and melting points (cryometry; e.g., FABRICIUS 1984).
   1.2 Measurement of homogenization temperatures between liquid and gas phases in polyphase inclusions (thermometry; e.g., FABRICIUS 1984).
2. Direct methods (involving destruction of the inclusions).
   2.1 Mechanical destruction of many inclusions by grinding and subsequent extraction with water or organic solutions (ROEDDER 1958; KRAMER 1965; PETRICHENKO 1973: 246 ff).

2.2  Release of the solutions from many inclusions by heating and melting in a vacuum (decripitation methods) used for determining $\delta^{18}O$ and $\delta D$.

2.3  Extraction of the solution from *individual* inclusions (e.g., HOLSER 1963, 1968; PETRICHENKO 1973: 242 ff; LAZAR & HOLLAND 1988; HERRMANN et al. 1991). Fluid inclusions with diameters > 250 µm are opened with a microdrill and extracted using a specially designed micropipette (LAZAR & HOLLAND 1988; DAS et al. 1990; HERRMANN & v. BORSTEL 1991; HERRMANN et al. 1991; HORITA et al. 1991; v. BORSTEL 1992).

With microthermometric methods only the salinity and the ratios of several ions in the fluid inclusions can be estimated, but not the quantitative amounts of major and trace elements.

The solutions occurring in the evaporites, as well as the fluid inclusions in halite (e.g., LAZAR & HOLLAND 1988; v. BORSTEL 1991; HERRMANN & v. BORSTEL 1991), can only be described with quaternary (e.g., $NaCl-KCl-MgCl_2-H_2O$) and quinary (e.g., $NaCl-KCl-MgCl_2-CaCl_2-H_2O$, $NaCl-KCl-MgCl_2-Na_2SO_4-H_2O$) systems. Consequently, the genesis and origin of such solutions can only be interpreted when the major elements Na, K, Mg, Ca, Cl, and $SO_4$ and the trace elements Li, Sr, and Br can be determined with sufficient accuracy (e.g., HOLSER 1963, 1968; PETRICHENKO 1973; ROEDDER 1984: 432; LAZAR & HOLLAND 1988; HERRMANN & v. BORSTEL 1991; HERRMANN et al. 1991; v. BORSTEL 1992). The measurement of pH, Eh, and hydrogen-oxygen isotope ratio are also necessary.

The best method currently known for the quantitative determination of the major and trace chemical constituents in fluid inclusions was developed and described by LAZAR & HOLLAND (1988). This method involves an optimized extraction technique which allows the determination of the aforementioned elements by means of ion chromatography with relatively little experimental and instrumental expenditure. The incorporation of data on fluid inclusions in salt minerals in the quantitative interpretation of evaporite-forming processes first appeared meaningful following these advances in the development of microanalysis techniques with reproducibilities between 2% and 4% (LAZAR & HOLLAND 1988; HERRMANN & v. BORSTEL 1991; HERRMANN et al. 1991; v. BORSTEL 1992).

Of the different ion chromatographic methods, particularly the ultramicroanalytical techniques (e.g., PETRICHENKO 1973), atomic emission spectroscopy with inductively coupled plasma (ICP-AES; e.g., ROEDDER 1984), and Raman laser spectroscopy (e.g., STEIN 1985) have been employed.

Previous scientific studies have dealt primarily with questions concerning the compositional evolution of seawater in the geological past. Objects of study have mainly been rock salts which formed over the last 900 Ma (KRAMER 1965; LAZAR & HOLLAND 1988; DAS et al. 1989; HORITA et al. 1991). It has been determined that the composition of seawater in the past was basically not different from that of today's oceans.

Furthermore, there are indications of lower amounts of $SO_4$ in fluid inclusions from rock salt crystals (e.g., PETRICHENKO 1973; ROEDDER 1984; v. BORSTEL 1992), which is indicative of $MgSO_4$ depletion in salt solutions during evaporite formation (HERRMANN 1991c).

Russian authors offer evidence that metamorphic processes should be considered when interpreting the composition of fluid inclusions in salt minerals. For example, due to homogenization temperatures of about 120 °C for polyphase inclusions (liquid-gas) in the rock salt of salt domes of the Dnjepr-Donez depression (Upper Devonian) the effect of thermal metamorphism is being taken into consideration (e.g., PETRICHENKO 1973: 263). Russian authors also point out the possibilities for applying fluid inclusion studies in the prospection of potash salt occurrences (PETRICHENKO 1973: 260).

Some data on the pH and Eh of solutions in the fluid inclusions of salt minerals have been collected. The pH values range from 2 to 11, averaging between 6 and 9 (PETRICHENKO 1973; 250). The Eh values vary between -260 and +320 mV. Solutions containing Fe(II) have lower Eh values (-260 to +100 mV) than those with Fe(III) whose Eh values indicate oxidizing behavior.

Information on the source of the aqueous solutions in the fluid inclusions and in the evaporite rocks may be obtained by determining the $\delta^{18}O$ and $\delta D$ values (e.g., ROEDDER 1984; O'NEIL et al. 1986; HORITA & MATSUO 1986; SCHMIEDL et al. 1982, 1983). Such studies are usually connected with problems in salt mining or the construction of underground repositories; in these cases the evaluation of the source of the water in the salt solutions plays an important role in safety analyses (e.g., O'NEIL et al. 1986). However, the interpretation of isotope data is difficult. In addition to the changes in isotope values during the evaporation of seawater, the fixation of $H_2O$ in hydrate minerals and mixing with formation and groundwaters must also be considered (e.g., ROEDDER 1984: 434). It was able to be demonstrated for fluid inclusions in the rock salt of the evaporites of the Delaware basin (Upper Permian, New Mexico and Texas) that mixtures of old and recent groundwaters are obviously widespread. $H_2O$ from the dehydration of gypsum is to be considered as well (O'NEIL et al. 1986).

What is the status of fluid inclusion studies in the salt minerals of the Zechstein evaporites in central and northern Germany?

Neither local nor regional systematic studies on the quantitative composition of fluid inclusions in the Zechstein evaporites have been conducted. Now that the analytical foundation has been laid in recent years by LAZAR & HOLLAND (1988), the research project described here should help close this gap (HERRMANN et al. 1991; HERRMANN & V. BORSTEL 1991; V. BORSTEL 1991). This project has two main aspects: (1) fractionation processes before and during evaporite formation, (2) fractionation processes during metamorphism and material transport in the evaporites after deposition of the crystallizate.

This means that in addition to studies of rock salt beds, more emphasis should be placed on research on the composition of fluid inclusions in the rocks of potash salt seams (rock salt, sylvinite, carnallitite) and in secondary minerals in fissures and cracks (halite, sylvite, carnallite, bischofite, kainite, kieserite, polyhalite) and in blue rock salt.

The literature cited in this chapter only serves as examples for the topics discussed and is by no means complete.

## 13.2 Research concept

The research concept described in the following should meet the prerequisites for a systematic study of the composition of fluid inclusions and salt solutions. This study will concentrate primarily on the evaporites of the Zechstein sequence 1-4 (HERRMANN 1987b).

The temporal and spatial evolution of a geological system must be considered in the calculation of material balances and the quantification of evaporite-forming processes. The following questions are important to the composition of the solid and liquid constituents of evaporite bodies:

*1. How great are the chemical differences and fractionations in the salt solutions from which the evaporites crystallized compared with present-day seawater composition?*

To obtain information on the possible composition of salt solutions in the depositional basin of evaporites in the geological past, fluid inclusions contained in the minerals of these evaporites must be studied. These inclusions should still be representative of the composition of the original crystallisates; i.e., their chemistry must have remained essentially unchanged during the evolution of the evaporite. For analytical purposes the inclusions must be at least 250 m in size.

Fluid inclusions in carnallitic rocks whose chemical and mineralogical composition still closely represent the original (primary) composition appear suitable for this purpose. In addition to the major elements, trace elements such as Br and Li are also important criteria.

The composition of fluid inclusions in halite crystals from the base of Na1 (the Werra rock salt) and Na2 (the Staßfurt rock salt) can also give indications as to the composition of the solutions of the former Zechstein basin. However, the following must be remembered:

a) Rock salt we observe today, whether flat-lying or steeply inclined, has certainly been extensively recrystallized over the last 250 Ma (with or without the participation of $H_2O$ from the dehydration of gypsum), during which the composition of the parent solutions also changed. Br is redistributed between solid and solution especially during slow recrystallization. A portion of the Br originally held in the halite structure passes into the solution phase; i.e., the halite crystal becomes poorer in Br. That is to say, the Br distribution coefficient for halite is dependent on the growth rate of the crystals. The distribution coefficient $Br_{NaCl}$ for the comparatively slow recrystallization are obviously lower than those for the faster NaCl crystallization from seawater (see, e.g., Herrmann 1972, 1980; HERRMANN et al. 1973; HERRMANN & v. BORSTEL 1991).

b) The sulfate-type evaporites of the Zechstein sequences 1 and 2 also crystallized from solutions depleted in $MgSO_4$ compared with recent seawater (HERRMANN 1991c). Hence, studies on fluid inclusions in minerals of Z1 and Z2 do not yield any direct information on the composition of Zechstein seawater. However, it should be possible to comment on fractionation processes which had already altered the composition of the Zechstein sea in the deposional basin of the salt before the evaporite crystallized.

## 2. What is the composition of the solutions involved in the mineral reactions (diagenesis, metamorphism)?

Does the composition of the evaporite remain unchanged from the time of crystallization from seawater up to the present? In other words, do the various carbonate, sulfate, and chloride rocks remain in an isochemical state following their crystallation from concentrated seawater in the Upper Permian 250 million years ago?

In light of these questions, the study of the composition of the solutions during metamorphism is of key interest. There are clear indications, above all in the Thüringen, Hessen, Staßfurt, and Ronnenberg potash salt seams of Zechstein sequences 1-3, that following their formation the chemical and mineralogical composition of the original mineral associations have been altered by unsaturated aqueous solutions. This is equally true for the flat-lying evaporite beds of Z1 in the Werra-Fulda district and the steeply inclined salt beds in the domes of central and northern Germany.

Fluid inclusions in the altered mineral associations of the rocks Hartsalz, sylvinite, carnallitite, blue sylvinite (composed of halite, sylvite, carnallite, kieserite?, kainite) are suitable for study. The so-called Hemelinger carnallite (potash seam Ronnenberg, Salzdetfurth mine) or the bischofite from Vienenburg are of interest as well. The bischofite of Vienenburg has been genetically interpreted by HERRMANN (in BRAITSCH 1971: 201).

By studying the fluid inclusions in secondary carnallite (e.g., Hemelinger carnallite from Salzdetfurth) decisive knowledge on the genesis of highly concentrated $MgCl_2$ solutions could possibly be obtained. For example, bischofite-saturated fluid inclusions in halite and carnallite crystals as well as the contents of Br and Li would be important criteria (see HERRMANN & v. BORSTEL 1991).

The trace element Br plays an essential role in the context of the aforementioned problems for the following reason: In contrast to halite, the Br distribution coefficients for sylvite, carnallite, and bischofite are practically independent of the growth rate of the crystals. Thus, for fluid inclusions in primary or secondary sylvite, carnallite, and bischofite there is a clear genetic relationship between the amount of Br in the mineral and in the fluid inclusions.

The Li content, which is also to be determined with ion chromatography, is of interest with respect to information on the genesis and source of aqueous solutions involved in the mineral reactions (seawater, formation waters, magmatogenic waters, surface waters, water of crystallization during thermal metamorphism).

Studies of the composition of fluid inclusions in salt minerals can also be used to check whether the experimentally determined solution equilibria, which serve as the basis for the quantitative genetic interpretation of evaporite-forming processes, agree with the actually observed solution compositions. For example, it would be interesting to know whether the natural solutions are supersaturated with certain elements or compounds compared with solutions with stable equilibria.

## 3. To what extent have solutions migrated into the evaporites?

To answer this question, the secondary mineralizations in cracks, fissures, and caverns (crystallization with decreasing temperature) must be given the same consideration as mineral recrystallizations in areas of salt wash surfaces and salt tables. Minerals of primary interest with regard to fluid inclusions are halite, sylvite, carnallite, bischofite, kainite, and polyhalite.

In evaporite rocks (especially chlorides) the comparatively low porosity and permeability as paths for mobile phases do not play as decisive a role as fractures in the form of cracks and fissures. The latter form due to mechanical stress on an evaporite body, primarily in anhydrite beds, but also in rock salt, salt clay, and potash salt rocks. Numerous observations of this have been made both in flat-lying and steeply inclined evaporite bodies (e.g., WEIß 1980).

The composition of the minerals found in fissures and cracks frequently cannot be inferred from the surrounding rock. Rather, the minerals in dissolved form have been transported over great distances to the present site. Examples are $MgCl_2$-, $KCl$-, and $NaCl$-rich solutions found in anhydrite and rock salt beds (e.g., the Leine anhydrite A3, Staßfurt rock salt Na2, and Leine rock salt Na3 in the salt domes of northern Germany). Little is known about the distances that these solutions traveled or the transport mechanism. Yet with regard to underground repositories for anthropogenic wastes, questions as to transport distances and mechanisms are becoming increasingly relevant (e.g., HERRMANN 1984a; HERRMANN & KNIPPING 1989; see also Section III).

Investigations of the oxygen and hydrogen isotope ratios allow conclusions to be drawn on the source of the salt-bearing waters. However, the genetic interpretation of the measurements is often difficult (see Chapter 13.1).

Studies of the Li content are also important in gathering information on the source of salt solutions. Much of our own unpublished data on salt solutions from cracks, fissures, and caverns in evaporite rocks show that Li contents over 30 µg Li/g solution are most probably related to formation waters from the rock surrounding the evaporite bodies (HERRMANN). In contrast, solutions involved in metamorphism whose composition originates entirely from the dissolution of evaporite rock by surface waters only contain about 2 μg Li/g solution. This is due to the fact that Li is not fixed in the crystallizates during the various stages of seawater evaporation, but remains in the residual solutions (unpublished studies). These observations should be confirmed with fluid inclusion studies.

Through the study of fluid inclusion composition (major elements, trace elements, $\delta^{18}O$, $\delta D$, pH, and Eh) it should be possible to get an idea of the extent of solution transport in the evaporites. Information of this type is not only of great value to fundamental research on evaporite-forming processes, but also to statements on the long-term safety of underground repositories. Therefore, this aspect will also be considered in this research project.

# 14 Gases in marine evaporites

## 14.1 Fundamentals

In addition to the solids (minerals, rocks), marine evaporites contain varying amounts of liquids (salt solutions, rare condensates and oil) and gas mixtures, which under certain conditions can physically and chemically interact. This is true regardless of the age of the evaporites (i.e., of whether they formed in the Paleozoic, Mesozoic, or Cenozoic) and their geographical distribution.

Previous studies of evaporite deposits have concentrated primarily on the solid constituents and the local occurrence of salt solutions. Another area of evaporite research of equal importance to fundamental research, mining, underground storage, and underground disposal deals with local occurrences of salt-related gases.

The gases contained in evaporites are nearly always mixtures of several elements and compounds. Particularly the composition and volume of the gases trapped in evaporites vary regionally. These factors influence in various ways the mining of gas-bearing evaporites as well as the construction of underground repositories in evaporites and the evaluation of the long-term safety of repositories.

Important knowledge on the occurrence of gases in marine evaporites has been gained over the past 125 years, primarily from studies of the Zechstein evaporites in central and northern Germany. No other salt deposits in the world have been studied and documented in such detail. Consequently, the Zechstein evaporites of Germany will be the primary reference for this study.

Compared with other salt occurrences in the world, the Zechstein evaporites underlying central and northern Germany are very unique with respect to their solid, liquid, and gaseous components. Zechstein sequences 1 and 2 belong to the 0.5-5 % of all evaporites classified as sulfate type and 99.5-95 % of all salt deposits are chloride type (HERRMANN 1991c).

The salt solutions contained in the Zechstein evaporites of central and nothern Germany are distinguished by the range of variation in their composition, concentration, formation, and origin (e.g., BAUMERT 1928, 1952; HERRMANN 1961a, 1961b; SCHWANDT 1973/74; v. BORSTEL 1992). The gases detected in the Zechstein deposits are also characterized by their varying composition, formation, and volume. Large quantities of dominantly $CO_2$-bearing gases penetrated into the 250-million years-old evaporites of Zechstein 1 during the intrusion of basaltic melts into the Werra-Fulda mining district 16 - 18 million years ago. In contrast, hydrocarbons and nitrogen of other origin dominate the gases contained in rocks of Zechstein sequences 2, 3, and 4 of central and northern Germany. Isolated occurrences of liquid hydrocarbons (condensates, petroleum) have also been detected in evaporites.

With respect to the underground disposal of anthropogenic wastes in horizontal or steeply inclined Zechstein evaporites, the gases contained therein must be taken into consideration in two ways: firstly, by constructing underground repositories with mining or solution techniques, and secondly, when evaluating the long-term safety of underground repositories.

The site-specific geology and mineralogy of repository sites are naturally the most important factors. However, they can only be evaluated reliably by comparisons with gas occurrences in rocks of other evaporite deposits. Such comparative studies involve, for example, the determination of the composition, volume, mobility, and fixation of gases in evaporites under certain geological and mineralogical conditions. Although the composition and genesis of gases in evaporites of the Werra-Fulda mining district differ from those of the gases in salt domes in Niedersachsen (Germany), the type of gas inclusions and the possible transport of gas by accompanying aqueous solutions does not basically differ between the two regions. Based on observations of the juxtaposition of solutions and gases in cracks, fissures, and caverns in various evaporites (e.g., anhydrite, rock salt) there are obvious relationships between gases and salt solutions.

During construction of underground repositories, the general interaction between the solid, liquid, and gaseous components of evaporite occurrences are always to be noted carefully, in addition to the site-specific factors. This will be a primary consideration of this Part II. Therefore, the previous knowledge of gas occurrences in evaporites will first be summarized before a site-related scheme for the evaluation of gases in the Gorleben salt dome is developed. The results of surface exploration for the Gorleben salt dome became available as this Part II was written.

## 14.2 Gas occurrences in marine evaporites

Gases have been detected in many marine evaporites of the Paleozoic, Mesozoic, and Cenozoic. An initial introduction is given in the following works. This overview is by no means complete due to the poor availability of much information and unpublished data.

### *PALEOZOIC*

#### *Upper Devonian*

Pripjat depression, USSR:
KRASNOKUTSKI (1960), DUDYREV & SUNGUROVA (1963), after GIMM & ECKART (1968: 553).

## Lower Permian

Upper Kama, USSR:
IVANOV (1935: 9), SAVCHENKO (1958), KRASNOKUTSKI (1960), DUDYREV & SUNGUROVA (1963), the two last publications are after GIMM & ECKART (1968: 553, 572), BOL'SHAKOV (1972).
Caspi depression, USSR:
JEWENTOW et al. (1973)

## Upper Permian

Zechstein sequences in Germany:
REICHARDT (1860), SCHLEIDEN (1875: 157), OCHSENIUS (1877: 118), PRECHT (1879), FRANTZEN (1894), ERDMANN (1910), SCHEERER (1911), GROPP (1919), FIEGE (1934, which contains extensive references), LOTZE (1938, 1957), BAAR (1952), STOLLE (1953, 1954, 1974), HARTWIG (1954), SPACKELER (1957: 431ff), GIMM (1955, 1964), GIMM & ECKART (1968), ERLER (1957), MÜLLER (1958), JAHNE & PIELERT (1964), ELERT & FREUND (1969), KÄSTNER (1969), FREYER (1973, 1978), FREYER & WAGENER (1975), LANG (1973), MALZAHN (1973), HEMPEL (1974), EHRHARDT (1980), SLOTTA (1980), HORSTMANN & V. STRUENSEE (1981), ELERT & KNABE (1982), GRÜBLER (1983, 1984a, 1984b), ELERT & HENNING (1988), GIESEL et al. (1989), HEMMANN (1989), KNABE (1989), MIETH et al. (1989), SCHOTT (1989), GERLING et al. (1991).

Zechstein sequences in northern Jutland, Denmark:
DISPOSAL OF HIGH-LEVEL RADIOACTIVE WASTE (1981), GOMM (1982).

Zechstein sequences in the Kujawy region, Poland:
C. POBORSKI & J. POBORSKI (1964), after GIMM & ECKART (1968: 553, 572).


## MESOZOIC

### Triassic

Alpine evaporites, Germany:
SCHAUBERGER (1960).

### Upper Triassic to Lower Jurassic

Gulf Coast salt domes, USA:
HUNER (1939), BELCHIC (1960), MURRAY (1961: 209), HOY et al. (1962), KUPFER (1963: 117, 1980), THOMS & MARTINEZ (1980), MAHTAB (1981a, 1981b), IANNACCHIONE et al. (1984), SCHATZEL & HYMAN (1984).

### Cretaceous

Thailand

## CENOZOIC

### Tertiary

Carpatian fore-depression, western Galicia, Poland:
SCHLEIDEN (1875: 150), POBORSKI (1959), LITONSKI & BIALY (1964), after GIMM & ECKART (1968: 553, 572).
Upper Rhein Valley graben, France:
FIEGE (1934), VIGIER & PANCZUK (1961).


## 14.3   Inclusion of the gases

In marine evaporites gases are trapped in carbonate- and anhydrite rocks as well as rock and potash salts. Accumulations of gas can also occur in the immediate vicinity of salt clays.

The gases are fixed in the evaporites in the following ways:


### 1. Crack- and fissure-bound gases

Mixtures of gas (free gases) are trapped in fractures such as cracks, joints, and fissures. This type of gas fixation has already been described by NETTEKOVEN & GEINITZ (1905) for a body of salt in northern Germany, SCHEERER (1911: 214) and BECK (1912: 150) for evaporites of the Werra-Fulda mining district, and by BAAR (1952: 469) for the southern Harz Mtn. mining district. Various solutions together with the gases are also trapped in small cavernous voids (e.g., in the salt domes of Niedersachsen, BAUER 1991). According to HEMMANN (1989) and U. STÄUBERT & A. STÄUBERT (1989) the gases encountered in potash and rock salt mines are frequently stored in fissures.


### 2. Mineral-bound gases

Gases are trapped within and/or between mineral grains of salt rocks (e.g., NAUMANN 1911: 605; BAAR 1952: 471; PETRICHENKO 1973: 230). Accordingly, a differentiation is made between the two groups of mineral-bound gas defined as follows:

a. Incrystalline, ingranular (e.g., HARTWIG 1954: 14, 20), or intragranular (e.g., OELS-NER 1961: 8) bond. In this case, gases are enclosed in mineral grains in the form of microscopic bubbles or in round voids, frequently together with small amounts of saturated salt solutions. In thin section the incrystalline gases frequently appear as small, closely neighboring bubbles distributed chaotically through the crystal or arranged along crystallographic axes or planes (e.g., HARTWIG 1954: 21). The diameter of the gas inclusions ranges from > 100 µm to ≤ 1 µm (KÜHN 1951: 106; GÜNTHER, after GIMM 1964: 25). There are no indications that gaseous components such as $CO_2$ are also fixed in the crystal structure of the salt minerals (GÜNTHER, after GIMM 1964: 25).

b. Intercrystalline or intergranular bond (both terms used by HARTWIG 1954: 20). This involves the fixation of gases along the grain boundaries of minerals and in the voids in rocks, primarily in chloride- and chloride-sulfate-bearing rocks such as rock salt, sylvinite, carnallitite, and Hartsalz.

Up to 90 % of the mineral-bound $CO_2$ gas mixtures in the salt rocks of the Werra-Fulda mining district is assumed to be intercrystalline, and the remaining 10 % incrystalline (W. RICHTER 1953; GIESEL 1968: 103; HOFRICHTER 1976: 79). According to SCHALLER (in OELSNER 1961: 14) incrystalline gas made up 5.9 % of the total, while intercrystalline gas made up 94.1 % of the total in a gas-rich salt rock from the Menzengraben potash mine (Werra District).

The gas inclusions in salt minerals often contain portions of aqueous salt solutions, and more rarely liquid hydrocarbons (e.g., PETRICHENKO 1973: 230).

## 3. Adsorptive bonding?

In addition to crack-, fissure-, and mineral-bound gases, various authors still regard adsorptive bonding as an independent form of gas fixation in evaporites, especially with regard to the storage of $CO_2$ in the evaporites of Zechstein 1 in the Werra-Fulda mining district (e.g., MÜLLER 1958: 45; HOPPE 1960: 108). Adsorptive bonding is the ability of solids and crystalline compounds to store gas on their surfaces in the form of layered molecules. Since part of the crack-, fissure-, and mineral-bound gases are also fixed in part to mineral and rock surfaces in this manner (HARTWIG 1954: 22), an additional distinction between adsorptively fixed gases and crack-, fissure-, and mineral-bound gases will not be made.

In carbonates such as dolomite and limestone, gases are usually stored in pore space.

## 14.4 Composition of the gases

Most gas mixures in marine evaporites fall into three catagories (e.g., HEMPEL 1989):

1. Gas mixtures containing large amounts of combustible hydrocarbons ($CH_4$ and higher hydrocarbons) in addition to hydrogen, nitrogen, carbon dioxide, oxygen, noble gases, and hydrogen sulfide. With the appropriate composition and concentration such gas mixtures are easily ignited in mines by open fire or sparks.

2. Gas mixtures containing a large amount of carbon dioxide, in addition to nitrogen, hydrogen, noble gases, and hydrogen sulfide. High concentrations of such mixtures in a mine pose the danger of suffocation to workers.

3. Gases composed predominantly of nitrogen. An example is the gas released by Jüngeres Steinsalz (Na3) as a shaft was sunk in the Nordhäuser potash works in the southern Harz mining district (e.g., GROPP 1919: 37, 70; SPACKELER 1957: 433).

Extensive studies on the quantitative composition of gases trapped in evaporites have been conducted on the Zechstein evaporites in central Germany. SCHRADER et al. (1960, 1962), ACKERMANN et al. (1964), and KNABE (1989) developed methods for analyzing the gas-bearing evaporites from the Zechstein sequences of central Germany.

Tab. 6 - 9 contain data on the composition of various gas mixtures and their occurrence in the Zechstein evaporites of central Germany.

The gas mixtures in the Zechstein evaporites of central and northern Germany and in the evaporites of other countries normally contain the same components, but their volumetric proportions can vary greatly. With respect to flammable gas mixtures, for instance, the methane content may decrease, as the hydrogen content increases. Occurrence of such gas mixtures in carnallitic rocks of the Magdeburg-Halberstadt mining district are described, for example, by REICHHARDT (1860: 347), PRECHT (1879), ERDMANN (1910a), and GROPP (1919). SAVCHENKO (1958), for example, described the gases in carnallitite and sylvinite from the deposits of Solikamsk. According to ERDMANN (1910a) gas which escaped from the carnallitite of the Leopoldshall VI mine had the

**Tab. 6** Composition of hydrocarbon-bearing gas mixtures. Samples taken from the layered carnallitite, Trümmer-carnallitite, and sylvinite of the Staßfurt potash seam (K2) in the Bismarckshall potash mine near Bischofferode, southern Harz mining district. Analyses by HOFFMANN (1963; cited after R. SCHRADER, personal communication 1964; also in GIMM & ECKART 1968: 548); n.d., no data.

| Gases | No. of samples | Mean | Range |
|---|---|---|---|
| | | ml/100 g salt | ml/100 g salt |
| Total gas content | 53 | 4.4 | 0.6 - 9.5 |
| | | volume fraction in % | volume fraction in % |
| $CH_4$ | 15 | 20.0 | 5.0 - 40.0 |
| $CO_2$ | 15 | 50.0 | 10.0 - 75.0 |
| $N_2$ | 15 | 30.0 | 15.0 - 50.0 |
| $O_2$ | 4 | 0.8 | 0.3 - 1.5 |
| $H_2$ | 2 | 2.5 | n.d. |
| Ar | 3 | 0.25 | 0.2 - 0.3 |
| He | 1 | < 0.5 | n.d. |
| CO | 15 | < 0.05 | n.d. |
| higher hydrocarbons than $CH_4$ | 2 | < 0.1 | n.d. |
| $H_2S$ | 1 | < 0.1 | n.d. |

**Tab. 7** Composition of hydrocarbon-bearing gas mixtures from the Zechstein evaporites of firedamp-endangered potash districts in central Germany (from HEMPEL 1974: 592); n.d., no data.

| Gases | Volume fraction in % | |
|-------|-------------------|-------------------|
| | Fissure-bound gases | Mineral-bound gases |
| $CH_4$ | 10 - 60 | 5 - 40 |
| $CO_2$ | 0 - 5 | 10 - 75 |
| $N_2$ | 40 - 90 | 15 - 50 |
| $H_2$ | 0 - 10 | n.d. |
| higher hydrocarbons than $CH_4$ | 0 - 15 | < 1 |

**Tab. 8** Composition of primarily carbon dioxide-bearing gas mixtures. Drill core samples from the sylvinite of the upper accompanying bed of the Hessen potash seam (Sachsen-Weimar near Unterbreizbach, from ACKERMANN et al. 1964). For sampling and macroscopic examination of the samples see THOMA & ECKART (1964); n.d., no data.

| Gases | No. of samples | Mean | Range |
|-------|---------------|------|-------|
| | | ml/100 g salt | ml/100 g salt |
| Total gas content | 60 | 3.9 | 14.0 - 1.5 |
| | | volume fraction in % | volume fraction in % |
| $CO_2$ | 48 | 84.0 | 95.0 - 45.0 |
| $N_2$ | 47 | 14.0 | 50.0 - 4.0 |
| $CH_4$ | 47 | 1.0 | 9.5 - 0.1 |
| $O_2$ | 2 | 0.5 | 0.7 - 0.2 |
| $H_2$ | 1 | 0.4 | n.d. |
| Ar | 2 | 0.2 | 0.3 - 0.1 |
| He | 1 | < 0.5 | n.d. |
| CO | 47 | < 0.05* | n.d. |
| higher hydrocarbons than $CH_4$ | 2 | < 0.01* | n.d. |
| $H_2S$ | 3 | < 0.1** | n.d. |

*lower limit of detection  **$H_2S$ cannot be measured quantitatively

**Tab. 9** Composition of gases in rocks of Zechstein sequences 1 - 4. Mining districts in central Germany. From KNABE (1989: 359, 364) with recalculations of the total amount of gas from cm³/t rock into ml/100 g rock; n.d., no data.

| Stratigraphy | | Rock | No. of Samples | Average ml/100g | Volume fraction in % | | | |
|---|---|---|---|---|---|---|---|---|
| | | | | | $N_2$ | $H_2$ | $CO_2$ | $\Sigma$ HC |
| Z 4 | A4r | Grenz anhydrite | 2 | 0.15 | 93.9 | 5.2 | n.d. | 0.9 |
| | A4 | Aller anhydrite | 6 | 0.15 | 84.6 | 0.4 | 13.0 | 2.0 |
| | T4 | lower Aller clay | 6 | 0.14 | 93.7 | 5.5 | n.d. | 0.8 |
| Z 3 | Na3 | Leine rock salt | 42 | 0.03 | 98.4 | n.d. | n.d. | 1.6 |
| | Na3ß₂ | Liniensalz | 25 | 0.05 | 93.8 | n.d. | n.d. | 6.2 |
| | Na3ß₁ | Liniensalz | 10 | 0.03 | 93.4 | n.d. | n.d. | 6.6 |
| | Na3 | Schwaden- an.d. Tonflockensalz | 3 | 0.07 | 27.0 | 6.2 | 63.9 | 2.9 |
| | K3Ro | potash seam Ronnenberg, carnallitite | 10 | 0.13 | 90.5 | 4.3 | n.d. | 5.2 |
| | K3Ro | potash seam Ronnenberg, sylvinite | 19 | 0.21 | 80.6 | 11.0 | n.d. | 8.4 |
| | K3Ro | potash seam Ronnenberg, rock salt | 36 | 0.13 | 56.1 | 29.4 | n.d. | 14.5 |
| | A3 | Leine anhydrite | 6 | 0.39 | 30.9 | n.d. | 29.1 | 40.0 |
| | A3ε | Leine anhydrite | 10 | 0.24 | 34.5 | n.d. | 32.0 | 33.5 |
| | A3 | Leine anhydrite | 15 | 0.62 | 50.4 | 0.3 | 20.1 | 29.2 |
| | T3α | lower Leine clay, dolomitic | 48 | 0.65 | 43.3 | 1.2 | 26.6 | 28.8, |
| | T3ts | lower Leine clay, clayey-san.d.y | 31 | 0.09 | 86.5 | 11.0 | n.d. | 2.5 |
| | T3a | lower Leine clay, anhydritic | 29 | 0.05 | 84.4 | 11.5 | n.d. | 4.1 |
| | T3d | magnesite layer | 8 | 0.55 | 27.3 | 0.4 | 38.1 | 34.2 |
| Z 2 | A2r | upper Staßfurt anhydrite | 12 | 0.06 | 50.9 | 6.6 | n.d. | 42.6 |
| | K2 | potash seam Staßfurt, carnallitite | 69 | 0.51 | 71.8 | 10.1 | n.d. | 18.1 |
| | K2 | potash seam Staßfurt, Hartsalz | 13 | 1.1 | 91.3 | n.d. | n.d. | 8.7 |
| | Na2 | Staßfurt rock salt | 10 | 0.18 | 99.4 | n.d. | n.d. | 0.6 |
| | A2 | lower Staßfurt anhydrite | 17 | 0.28 | 77.6 | 1.9 | n.d. | 20.5 |
| | Ca2 | Staßfurt carbonate | 22 | 0.99 | 13.1 | 0.6 | 55.7 | 30.6 |

**Tab. 9** continued

| Stratigraphy | | Rock | No. of Samples | Average ml/100g | Volume fraction in % | | | |
|---|---|---|---|---|---|---|---|---|
| | | | | | $N_2$ | $H_2$ | $CO_2$ | $\Sigma$ HC |
| Z 1 | Na1γ | upper Werra rock salt | 12 | 0.06 | 79.2 | 12.8 | n.d. | 7.8 |
| | K1H | potash seam Hessen[1] | 79 | 0.16 | 84.3 | 7.8 | n.d. | 7.9 |
| | Na1ß | middle Werra rock salt[2] | 45 | 0.08 | 96.2 | n.d. | n.d. | 3.8 |
| | Na1ß$_4$ | middle Werra rock salt | 18 | 0.08 | 97.5 | n.d. | n.d. | 2.5 |
| | Na1ß$_3$ | middle Werra rock salt | 12 | 0.09 | 95.2 | n.d. | n.d. | 4.8 |
| | Na1ß$_2$ | middle Werra rock salt | 11 | 0.08 | 95.9 | n.d. | n.d. | 4.1 |
| | Na1ß$_1$ | middle Werra rock salt | 4 | 0.25 | 58.9 | 25.7 | n.d. | 15.4 |
| | K1Th | potash seam Thüringen[1] | 81 | 0.18 | 68.1 | 21.3 | n.d. | 10.6 |
| | Na1α | lower Werra rock salt[3] | 45 | 0.08 | 98.6 | n.d. | n.d. | 1.4 |
| | Na1α$_4$ | lower Werra rock salt | 7 | 0.09 | 98.6 | n.d. | n.d. | 1.4 |
| | Na1α$_3$ | lower Werra rock salt | 10 | 0.06 | 99.2 | n.d. | n.d. | 0.8 |
| | Na1α$_2$ | lower Werra rock salt | 19 | 0.09 | 98.6 | n.d. | n.d. | 1.4 |
| | Na1α$_1$ | lower Werra rock salt | 9 | 0.09 | 97.9 | n.d. | n.d. | 2.1 |
| | A1α | lower Werra anhydrite | 10 | 0.25 | 61.0 | n.d. | n.d. | 39.0 |
| | CaA1 | Anhydritknoten-schiefer | 2 | 0.26 | 7.4? | 1.2 | 45.9 | 45.5 |
| | Ca1 | Werra carbonate | 5 | 0.89 | 3.0 | 0.3 | 93.2 | 3.5 |
| | T1 | lower Werra clay (Kupferschiefer) | - | n.d. | n.d. | n.d. | n.d. | n.d. |
| ro | | Rotliegen.d. | 2 | 0.87 | 16.4 | 1.5 | 81.4 | 0.6 |

[1]average    [2]average without Na1ß$_1$    [3]average Na1α

following composition: 83.6 vol% $H_2$, 4.4 vol% $CH_4$, 12.0 vol% residual gas (e.g., $N_2$, He, Ne). However, PANETH & PETERS (1928b: 191) point out that the occurrence of helium and neon may be caused by contamination of the gas sample with air. The gas escaping from the carnallitite of the Leopoldshall VI mine was ignited by a blast and produced a m-long flame from the borehole (ERDMANN 1910). Hydrogenous gas mixtures occur also in the upper parts of the potash salt horizons (carnallitite and also sylvinite) in the deposits of the upper Kama as well (Solikamsk, Beresniki; e.g., SAVCHENKO 1958: 19). According to data of GIMM & ECKART (1968: 533, 572), these gas mixtures consist of 0-27 % $CH_4$, 0-46 % $H_2$, 34-97 % $N_2$, and 0-5.3 % $CO_2$. Similar compositions are yielded from the data of CHEREPENNIKOV (in SAVCHENKO 1958: 20) for the gas-bearing carnallitite and sylvinite from Solikamsk. $CO_2$-bearing gas mixtures are found in the salt domes of Louisiana (Gulf Coast, USA; HOY et al. 1962: 1458; KUPFER 1980: 122). An analysis by J. SAUNDERS (in HOY et al. 1962: 1458) yielded the following composition: 49.6 vol% $CO_2$, 17.3 vol% $H_2O$, 18.4 vol% $N_2$, 4.8 vol% CO, 4.4 vol% $O_2$, 3.7 vol% $SO_2$, 1.8 vol% $H_2$, 1.5 vol% $CH_4$, 0.4 vol% Ar, 0.4 vol% $C_2H_2$, and 0.4 vol% other hydrocarbons. Gases and moisture have frequently been found to occur together in the salt domes of the Gulf Coast (e.g., THOMS & MARTINEZ 1980).

Analytical results of gas mixtures from evaporites must be evaluated critically. For example, according to TAMMANN & SEIDEL (1932) the gas inclusions in 'knister' (crackle) salt from the Sachsen-Weimar potash mine near Unterbreizbach contain mostly air with the following composition: 75.7 vol% $N_2$, 15.3 vol% $O_2$, 5.8 vol% hydrocarbons, 3.3 vol% $CO_2$. However, it is very doubtful that this analysis yielded a representative composition. TAMMANN & SEIDEL (1932) investigated knister salt by dissolving it in destilled water and analyzing the escaping gas. By doing so, a great portion of the $CO_2$ released was probably dissolved in the water: Hence, the $CO_2$ contents in the gas analysis would be too low (see also KÜHN 1951: 106). The analytical results in Tab. 8 for the same Sachsen-Weimar potash mine and numerous observations of gas releases in the Werra-Fulda mining district clearly indicate that the gas in knister salt consists of $CO_2$-bearing mixtures and do not have the compositon of air.

The results of FREYER (1973, 1978) and FREYER & WAGENER (1975) are to be regarded with caution as well. They assume that the gases in gas-poor evaporites in their studies (about 1 ml gas/kg carnallitite and Hartsalz from the Thüringen and Hessen potash salt seams, Werra rock salt Na1ß; all samples from the Hattorf mine) contain atmospheric components ($O_2$, $N_2$, $CO_2$, Ar), which supposedly entered these rocks during formation of the evaporites or recrystallization of the minerals. Numerous observations of the metamorphism of saline mineral associations of the last 250 million years do not support this hypothesis.

Using recent $^{13}C$ determinations (-7 ‰) KNIPPING (1989) was able to prove that the $CO_2$ in the gas-bearing evaporites of the Werra-Fulda mining district are directly related to the basalt magmatism 14 - 25 million years ago.

There are obviously analytical uncertainties in the data of MÜLLER & HEYMEL (1956; see SCHRADER et al. 1969).

Another source of error in the study of salt-bound gases is due to the fact that during sampling air finds its way into the gas mixtures contained in the evaporites. A correction for this air based on the oxygen content in the sample surely does not essentially improve the quality of the gas analysis (e.g., BAAR 1954: 341).

KNABE (1989) presented interesting results from his studies on the occurrence and quantitative composition of gases in the evaporites of Zechstein sequences 1 - 4 (Tab. 9). The following observations are of interest here: In Zechstein sequences 1 - 4 the absolute amounts of gas contained in the evaporites decrease in the following order: carbonates, calcium sulfate, chloride rocks (Tab. 9). In potash seams this is only true for rocks of normal character (KNABE 1989: 360). This means that these evaporites have remained practically unaltered with respect to their original composition. The greatest amounts of gas (0.8 - 1.1 ml/100 g rock) were found in the Werra carbonate (Ca1, Zechsteinkalk) and in the Staßfurt carbonate (Ca2, Hauptdolomite). The gas contents in lithologically similar beds increase in the carbonate and chloride rocks from Z1 to Z2 and decrease again toward Z4. In contrast, the gas contents in the anhydrites rocks showed no distinct trend through the Zechstein sequences.

The Werra, Leine, and Aller rock salt (Na1, Na3, Na4) contain the lowest gas contents (Tab. 9). Hydrogen occurs above all in the carbonate and calcium sulfate rocks but is rare in rock salt. The potash salt seams are not included here (KNABE 1989: 358). Nitrogen dominates in the rock salt beds, while $CO_2$ and hydrocarbons are fixed predominantly in carbonates and anhydrite beds.

It is noteworthy that the rocks of normal character from the Thüringen (K1Th) and Hessen (K1H) potash seams in the Werra-Fulda mining district contain no $CO_2$ (KNABE 1989; Tab. 9). In Zechstein sequence 1, however, the rocks of both potash seams contain distinctly greater amounts of hydrogen, also accompanied by greater amounts of hydrocarbons. The highest gas contents were found in the secondary sylvinite of the K1Th and K1H potash seams (KNABE 1989: 360).

The investigations of KNABE (1989) also clearly demonstrate that the gas-rich and $CO_2$-rich rocks of Zechstein sequence 1 of the Werra-Fulda mining district are undoubtedly the product of alteration during which fluid phases penetrated the evaporites and gas mixtures permeated the salts (see also KNIPPING 1989).

## 14.5 Gas volumes

Data on the volume of gases contained in various evaporites are available in the scientific literature, above all on the Zechstein evaporites of central Germany (Tab. 6, 8, 9). Obtaining such data involves problems which can best be explained using the gas-rich evaporites of the Werra-Fulda mining district as an example.

Fourteen gas explosions occurred between 1908 and 1957 at the level of the Thüringen and Hessen potash salt seams. Following these large gas outbursts, the mass of displaced salt rock and volume of gas released were estimated for the $CO_2$-bearing gas mixtures in the evaporites of the following potash mines: Menzengraben, Sachsen-Weimar, Kaiseroda, and Dietlas (eastern part of the Werra-Fulda mining district; e.g., LIEBSCHER 1952; MÜLLER 1958: 34; DUCHROW 1959). Considering all 14 explosions, the evaporites contained between 3 and 25 $m^3$ gas/t salt. The average value totals 14.6 $m^3$ gas/t (see also ECKART et al. 1966; WOLF cited in GIESEL et al. 1989). Consequently, these values are one to two orders of magnitude higher than the experimentally determined volumes of 0.04-0.6 $m^3$ gas/t salt in the rocks of the same potash mines (OELSNER 1961; 14; see also Tab. 8). For gas-bearing evaporites from the Staßfurt (K2) potash salt seam of the southern Harz mining district HEYMEL (in GIMM 1954: 592) calculated 0.1 $m^3$ gas/t rock in the Bleicherode potash works and up to 0.16 $m^3$ gas/t rock in the Bismarckshall potash works. In December 1973 there was a great outburst of mixed gas containing nitrogen and hydrocarbons in the very gas endangered Bischofferode potash mine, which created a void of 10 000 $m^3$. The volume of gas released per ton of displaced salt is estimated to be 3-5 $m^3$ (HEMPEL 1974). Unfortunately, HEMPEL did not specify at which stratigraphic level the outburst occurred.

According to ELERT & KNABE (1982) and KNABE (1989) the Hartsalz in the Staßfurt potash salt seam of the southern Harz and Unstrut-Saale mining districts contains 0.01 $m^3$ gas/t rock, and the carnallitite 0.005 $m^3$ gas/t rock.

In the Braunschweig-Lüneburg rock salt mine (Niedersachsen, Aller valley) gas-bearing horizons were degassed by drilling. In this way one million cubic meters of gas (nitrogen-methane) were released above ground (EHRHARDT 1980).

In Solikamsk (former USSR) carnallitite and sylvinite contained about 0.01 $m^3$ gas/t, of which 0.002 $m^3$ was hydrogen (CHEREPENNIKOV, in SAVCHENKO 1958: 20).

In rock salt of the Louisiana salt domes (Golf Coast, USA) release rates of 0.4-1.8 $m^3$ $CH_4$/t rock were measured in outburst zones. Normal rock salt contains less than 0.1 $m^3$ $CH_4$/t rock (IANNACCHIONE et al. 1984). There have also been sudden gas outbursts in the course of mining in salt domes of Louisiana, releasing great volumes of gas and ejecting up to several thousand metric tons of salt rock. For example, on 8 June 1979 in the Belle Isle mine there was a great gas outburst (mainly $CH_4$) at about 400 m depth, which released about 110 000 $m^3$ of gas (according to MAHTAB 1981a). The voids created by the outbursts have diameters of 1-30 m and lengths of 1-50 m and more (after MAHTAB 1981b; THOMS & MARTINEZ 1980).

The great differences between the amounts of gas inferred from the gas outbursts and those determined experimentally for the evaporites of the Werra-Fulda mining district cannot be explained by methodical errors alone (see also GIMM 1964: 23). Thus, ACKERMANN et al. (1964; 677) do not rule out the possibility that the gas-rich evaporites lost considerable amounts of gas during sampling. Gas can also be lost from the samples during storage, comparable with the partial degassing of evaporites when pressure is relieved or under the influence of overburdon pressure (e.g., MÜLLER 1958: 20, 63). In addition, the gas in cracks and fissures cannot be quantitatively determined during drill core sampling. However, it is possible that the portion of gas in cracks and fissures participating in a large gas outburst is considerable (e.g., BAAR 1977: 168). In addition, the portion of intercrystalline gas (about 90 %) is obviously greater than the portion of incrystalline gas (about 10 %; see Chapter 14.3).

Finally, it is also uncertain whether and to what extent accumulations of liquid $CO_2$ were also released by the large gas outburst in the Fulda-Werra mining district. Using microscopy OELSNER (1961: 15) found that incrystalline $CO_2$ is liquid at room temperature. In our opinion, however, this still needs to be confirmed by further study. The gas possibly occurs with salt solutions in the small bubbles. V.S. SHAYDETS-KIY (cited in PETRICHENKO 1973: 214) also observed liquid $CO_2$ as inclusions in halite from salt domes in the Dnjepr-Donez depression.

For these reasons it is just not possible to compare the amounts of gas released by an outburst per ton of evaporite with the volumes determined experimentally. In an outburst gases trapped in cracks and fissures and, in part, minerals are released. The volumes determined experimentally are primarily those of gases trapped in minerals.

It follows from the comparison between the volumes of gas per metric ton of evaporite estimated from gas outbursts and those experimentally calculated that obtaining representative data on the gases contained in evaporites obviously still involves experimental problems. This should always be remembered when evaluating data on the amount of gas in an evaporite rock.

## 14.6 Gas pressures

There are various data on the pressures exerted by mineral-bound gases (incrystalline, intercrystalline fixation). Based on the investigations of TAMMANN & SEIDEL (1932: 215) the gas in the knister salt of the Sachsen-Weimar potash works near Unterbreiz-bach (Werra mining district) is under a pressure of 9.8 - 11.8 bar or 0.98 - 1.2 MPa. KÜHN (1951: 106) calculated pressures of 7.8 - 9.8 bar or 0.78 - 0.98 MPa for the incrystalline gas in knister salt from the Hattorf mine.

After boreholes in the gas-bearing evaporites of Werra sequence were sealed, pressures of 20.6 - 24.5 bar (2.1 - 2.5 MPa) were measured (MÜLLER 1958: 20).

According to GIESEL (1968: 103) the gases enclosed in the evaporites of the Werra-Fulda mining district are apparently in a liquid state at a pressure of about 100 bar (10 MPa).

Maximum pressures of 100 bar (10 MPa) were also measured for inflammable gases trapped in fissures in the Zechstein evaporites in potash mines of central Germany (HEMPEL 1974). The pressure of the gases exerted in the fissures and cracks of evaporites can exceed significantly the hydrostatic pressures (HEMPEL et al. 1981).

Gas pressures of up to 60 bar (6 MPa) were measured at the head during underground drilling in the Braunschweig-Lüneburg mine (Aller valley; EHRHARDT 1980).

During exploratory drilling in the Leine rock salt pressures ranging from 8.4 to 13 MPa were measured in a zone of anhydrite swelling in the Bernburg-Gröna rock salt mine. These high gas pressures are obviously caused by the small volumes of the voids (150 l), i.e., the pressure had not yet been able to dissipate into the surrounding rock (HEMMANN 1989: 394).

The gases in 50 fluid inclusions in rock salt crystals from Zechstein evaporites in central and northern Germany were at pressures of < 2 to 25 MPa (v. BORSTEL 1991).

In the deposits of the Upper Devonian Pripjat depression (Soligorsk, former USSR) and at the upper Kama (Lower Permian) pressures up to 50 bar (5 MPa) and more were measured for the free gases occurring in the carnallitite or clayey carnallitic rocks (KOWALOW 1981). According to PETRICHENKO (1973: 230) gas trapped in salt minerals is under a pressure of several hundred bar.

According to BAAR (1977: 169) gas at a depth of 800 m is under a petrostatic pressure of about 200 bar (20 MPa). This means that 200 $m^3$ of a gas mixture under normal pressure would only have a volume of 1 $m^3$ under a pressure of 200 bar and at the same temperature ($p \cdot V$ = constant).

According to HOY et al. (1962: 1458) $CO_2$-bearing gas mixtures in the knister salt (crackle salt) of the Winnfield salt dome (Louisiana, USA) is under a pressure of 490 - 980 bar (49 - 98 MPa) at 0°C. These extremely high pressures were deduced from the estimated volume of gas releases and the volume of the small, gas-filled voids (bubbles) in the salt crystals. Similar data (500 - 1000 bar or 50 - 100 MPa) for gas bubbles in rock salt of a Louisiana salt dome are given in HYMAN (1982). During exploratory drilling in a Louisiana salt dome gas was released under a pressure of 62 bar (6.2 MPa) at a flow rate of 1.2 $m^3$/h (IANNACCHIONE et al. 1984).

## 14.7 Formation and origin of the gases

### 14.7.1 Fundamentals

Seawater also contains dissolved gases, e.g., 15 500 g $N_2$/l and 6 000 g $O_2$/l (mean values; KÖNIG et al. 1964; SVERDRUP et al. 1942 quoted according to TUREKIAN 1969: 309). In the following example it is assumed that nitrogen and oxygen accumulates in evaporating seawater and is subsequently fixed as gas in rock salt. To form one metric ton of rock salt about 39 metric tons of seawater must evaporate at 25°C until it is saturated with bloedite (evaporation of 1000 g of seawater to 25 g of solution, entry into the quinary system). It is also assumed that all nitrogen and oxygen originally contained in the 39 metric tons of seawater is contained entirely in one metric ton of rock salt. At 25°C and under a pressure of 1 bar that is about 0.53 $m^3$ $N_2$/t and 0.18 $m^3$ $O_2$/t of rock salt. The orders of magnitude of these gas volumes are nearly comparable with the amounts of gas determined experimentally in evaporites from the southern Harz and Werra-Fulda mining districts (Chapter 14.4, Tab. 6 and 8). Yet, can this be used as evidence that the gases fixed in evaporites are derived directly from seawater? This is surely not always the case.

The equilibrium calculation above is based on the assumption that the gases originally dissolved in the seawater remain in the salt solution becoming more and more concentrated until the composition of the solution enters the quinary system (i.e., crystallization of bloedite). However, the portion of molecularly dissolved gases does decrease in concentrated salt solutions. According to KINSMAN et al. (1974) only 1000 - 3000 g of dissolved oxygen per kg $H_2O$ (or 735 - 2205 g $O_2$/kg NaCl solution) are left in NaCl-saturated solutions. This converts to 0.02 - 0.07 $m^3$ $O_2$/t or 20-70 $cm^3$ $O_2$/ kg rock salt. The portion of molecularly dissolved nitrogen probably also decreases with increasing salt concentration in a similar fashion. It is also to be considered that the proportions of the various gases dissolved in seawater differ from those of the gas mixtures in evaporites. In short, the fact that differing amounts of gases are stored in evaporites of differing deposits as well as many observations of gas and solution migration in evaporites allow the conclusion to be drawn that there is no direct genetic connection between the formation of gas-bearing evaporites and the gases dissolved in seawater.

FREYER (1973, 1978) and FREYER & WAGENER (1975) assume that gases of the Zechstein atmosphere ($O_2$, $N_2$, $CO_2$, Ar; on the order of 1 ml/ kg rock) are contained in the evaporites of Zechstein 1 in the Werra-Fulda mining district. It seems unlikely, however, that a 250-million years-old atmosphere is preserved considering the metamorphism these rocks experienced. Without going into detail, it is pointed out that salt minerals crystallize not only at the surface of a saturated body of water but can also form in deeper water strata due to the temperature gradient and on the bottom of the basin. Gas mixtures of atmospheric composition are certainly not available at depth for inclusion in crystals.

Studies of marine evaporites confirm again and again that primarily two processes are involved in the formation of gas-bearing evaporites (as already described, for instance, by GROPP 1919: 34):

1. The gases were formed or fixed during or after crystallization of the evaporite minerals: the prerequisite being the presence of substances (not seawater or atmospheric air!) from which gaseous components could have formed under certain reaction conditions. This applies above all, gas mixtures in which hydrocarbons and hydrogen dominate.

   At least some of the gases formed in this way were frequently stored in the same stratigraphic horizons and in genetically related rocks (carnallitite together with sylvinite or Hartsalz and rock salt within one potash salt seam; e.g., ELERT & FREUND 1969).

2. Following deposition of the evaporites, gases penetrated from an external source into the present-day resevoir rocks, where a portion of them were fixed. In this case the present reservoir rocks were not petrographically and stratigraphically identical with the rock and the location in which the gases formed. In the interest of genetic observations and mining problems the following differentiations are made regarding the origin of the gases:

2a. The gases could have been formed and stored in the original deposits of an evaporite sequence (e.g., petroleum- and natural-gas-bearing carbonates, sulfate rocks in the form of anhydrite). From there, they migrated - often accompanied by solutions - by geological processes in the late-forming crystals of an evaporite sequence (rock salt, K-Mg minerals). This often involves hydrocarbon-bearing gas mixtures, e.g., the rocks of the Staßfurt sequence (Z2) of the southern Harz mining district.

Gases originally stored in anhydrite can migrate into and be fixed in chloride rocks such as carnallitite and rock salt due to changes in the temperature-pressure conditions and deformation of the evaporites. For example, SCHEERER (1911: 227) regards this as a possibility for the occurrence of hydrocarbon-bearing gas mixtures in the Straßfurt carnallitite from the Desdemona potash mine (upper Leine valley).

2b. The origin and formation of gases has nothing to do with the rocks of a evaporite sequence. Besides hydrocarbon-bearing gases, this applies above all to $CO_2$-bearing gas mixtures which penetrated into horizontal evaporite deposits as fluid phases accompanying basaltic melts where a portion of them were fixed. Examples are the rock salt and potash salt rocks of the Werra sequence (Z1) in the Werra-Fulda mining district.

2c. Gases whose affinity regarding origin and formation to the rocks of an evaporite sequence is speculative or unknown. The only thing sure about this group is that the gases penetrated into the salt rocks through fault zones from an external source (e.g., shear planes, etc.) and were fixed in the present-day reservoir rocks due to deformation of the evaporites (halokinesis or salt tectonics). Examples are some of the gases in rock salt of Gulf Coast salt domes, USA (e.g., BELCHIC 1960; HOY et al. 1962; KUPFER 1980) and of the Hope salt dome in Niedersachsen, Germany (HORSTMANN & v. STRUENSEE 1981: 58).

The gas mixtures in evaporites cannot always be assigned to a single genetic process. Thus, it is conceivable that a certain portion of the gases formed directly in the evaporites due to decomposition of organic substances. At a later time and under completely different geological conditions other foreign gases - some together with solutions - then penetrated into the same salt beds from an external source. However, making a clear distinction between gas mixtures which formed in the same way but have different sources is not easy. In favorable cases, indications of the source of the gases may be obtained from data on various hydrocarbons (e.g., AKSTINAT 1983) and isotope distributions (e.g., MAAß 1962; KÄSTNER 1964; SCHOELL 1983; SCHMITT 1987; GERLING et al. 1991).

How can gases form in evaporites?

## 14.7.2 Hydrocarbons

Hydrocarbons are derived from the decomposition of organic substances. The related processes most likely involve diagenesis (primarily microbiological activity) and catagenesis (increasing temperature with subsidence of sedimentary deposits). The connections between the occurrence of petroleum, natural gas, and marine evaporites are discussed in FIEGE (1934), BORCHERT (1940: 156ff, 1959: 151ff), LOTZE (1957: 377ff), BORCHERT & MUIR (1964: 232ff), ELERT & HENNING (1988), and ELERT et al. (1988). Petroleum and natural gas often accumulate along the margins of or near salt diapirs. In the USA such petroleum occurrences were already discovered during salt drilling in the first half of the 19th century (e.g., HALBOUTY 1979).

In Niedersachsen (Germany) the petroleum occurrence in the vicinity of the Hope salt dome is a geologically interesting locality. Petroleum occurrences are also known from areas of flat-lying salt sequences. For example, in 1930 in the Volkenroda potash mine (southern Harz mining district) petroleum and natural gas were released from the rocks below the potash salt seam. They originated from the original precipitates of the Staßfurt sequence ± 50 m below the Staßfurt rock salt (Na2) and the Staßfurt anhydrite (A2), from the Staßfurt carbonate or the Hauptdolomite (Ca2).

However, the occurrence of petroleum and natural gas in the marginal zones of evaporites does not always mean that liquid and gaseous hydrocarbons are to be expected within the salt rocks.

Organic matter is necessary for petroleum and natural gas to form. As organic matter is deposited in sedimentary environments with negative Eh (reducing conditions), they decompose, forming solid, liquid, and gaseous hydrocarbons (e.g., BRAITSCH 1962: 204, 1971: 263; KREJCI-GRAF 1962).

The development of an oxygen-poor or oxygen-free environment consisting of sediment and water and the formation of concentrated salt solutions necessary for evaporite deposition have something important in common. In both cases there is limited vertical mixing of the lower, negative-Eh and upper, positive-Eh water strata. Furthermore, there is very limited or no exchange between basin water and normal seawater from the open ocean. This means that the conditions for the development of a reducing environment are first met when the amount of water evaporating is greater than that being supplied to the basin (BORCHERT 1959: 151f; BORCHERT & MUIR 1964: 232). These reducing conditions can already develop before and during the deposition of the first precipitates of a saline sequence. All previous studies show that reducing conditions must have frequently existed during the subsequent crystallization of rock salt and K-Mg minerals (e.g., BORCHERT 1959: 153; BORCHERT & MUIR 1964: 233; BRAITSCH 1962: 205, 1971: 264). However, the iron in minerals such as carnallite and sylvite frequently occurs as hematite (e.g., BRAITSCH 1962: 172ff, 1971: 224ff; A. RICHTER 1962, 1964). In this case reducing conditions cannot have prevailed during the entire geological formation of these minerals and rocks.

How did the organic matter find its way into the reducing environment of the high-salinity basin? According to RIEDEL (1935: 72ff) the number of species already declines considerably when the concentration of seawater increases three to four times that of normal seawater. When seawater reaches about a 6-fold concentration (crystallization of gypsum, just before initial precipitation of halite), life is no longer possible. According to BORCHERT (1940: 159) this is already the case during initial gypsum crystallization. Even when life no longer exists in the concentrated seawater, it does appear likely that organic matter from dead micro-organisms is transported into the depositional basin by directed currents (e.g., LOTZE 1957: 379). In this way the conditions for the formation of more or less large quantities of hydrocarbons have been established for all the rocks of an evaporite sequence. The portion of total carbon is on the order of 1 % in the water-insoluble residues of the evaporites of the Zechstein of central and northern Germany. This is about double the amount of carbon in clays and shales (e.g., HOEFS 1969: 6-K-1) and corresponds to 0.01 % C with reference to the evaporite (BRAITSCH 1962: 204, 1971: 263). The bituminous smell that evaporites frequently have is also an indication of the presence of organic compounds.

ELERT & HENNING (1988) yielded around 20 - 1800 µg of bitumen A per g of rock from the Thüringen, Hessen, and Staßfurt potash seams and from the Staßfurt carbonate (Ca2) and lower Leine clay (T3). Cambrian rock salt in Siberia contains between 40 and 150 g of bitumen per g of rock (KONTOROVIĆ et al. after ELERT & HENNING 1988).

It is conceivable that part of this organic matter had already been deposited and finely dispersed during crystallization of the evaporite minerals (e.g., ELERT & FREUND 1969; ELERT & HENNING 1988). Part of the hydrocarbon-bearing gas mixtures in evaporites formed and was also fixed in the host rocks, i.e., rock salt, sylvinite, and carnallitite. Another and greater portion of the hydrocarbon-bearing gas mixtures and condensates probably penetrated into the evaporites from external sources, frequently in combination with aqueous solutions. An external source may be the rock enclosing the evaporite body or the initial precipitates of an evaporite sequence (e.g., carbonates). It is known that in the Zechstein deposits of the southern Harz mining district, aqueous solutions, gases, and petroleum have migrated from the gas- and petroleum-bearing initial precipitates of the Zechstein 2 salt sequence (Staßfurt carbonate Ca2, in part also Staßfurt anhydrite A2) up into the hanging rock layers of Zechstein 2 and the beds of Zechstein 3 (e.g., NAUMANN 1911: 618; ALBRECHT 1932; BAAR 1952, 1954; LIEBSCHER 1952). However, it is not sure whether the Staßfurt carbonate (Hauptdolomite) is both the source and reservoir rock for the liquid and gaseous hydrocarbons, or just the reservoir. The fact that there are also authigenic bitumens in the chloride rocks conflicts with the latter.

There are also minute quantities of gaseous and liquid hydrocarbons in the rocks of the Werra sequence (Z1) which are frequently related to basalt dikes, fracture systems, and the occurrence of $CO_2$-bearing gases (e.g., ELERT & HENNING 1988; ELERT et al. 1988). In many unaltered Zechstein evaporites (above all in chloride rocks) there is an obvious lack of $CO_2$ (see KNABE 1989; Tab. 9).

KÄSTNER (1964: 363) and ELERT et al. (1988) consider the possibility that hydrocarbons could have been mobilized and transported into the evaporites of the Werra sequence of Zechstein 1 during magmatic activity in the rocks underlying the Werra sequence. This would then be an example for the migration of hydrocarbon-bearing gas mixtures from rocks which genetically have nothing to do with the evaporites of the Werra sequence.

KNABE (1989) conducted extensive studies on the occurrence of hydrocarbons in the Zechstein evaporites of central Germany. Just as $CO_2$, hydrocarbons also dominate in carbonates and anhydrites (Tab. 9). The portion of the hydrocarbons ethane to n-hexane decreases significantly from the carbonates via anhydrite to the chloride rocks (KNABE 1989: 358). The Thüringen and Hessen potash salt seams contain greater amounts of hydrogen in addition to elevated hydrocarbon contents.

## 14.7.3 Carbon dioxide

The $CO_2$ in gas-bearing evaporites originates from two processes:

1. In the course of diagenesis and catagenesis $CO_2$ and $H_2O$ also form from organic matter (e.g., KREJCI-GRAF 1962: 21; SNARSKY 1963: 26; HUNT 1979: 168). Hence, at least a portion of the $CO_2$ occurring in the hydrocarbon-bearing gas mixtures can most likely be attributed to the alteration of primary organic substances.

2. The $CO_2$ in gas-bearing evaporites can also be of inorganic origin. In most cases the $CO_2$ is in fluid phases which penetrated into the evaporites during magmatic processes. Examples of the magmatic origin of $CO_2$-bearing gas mixtures are provided by the Zechstein 1 evaporites in the Werra-Fulda mining district (for recent experimental results see KNIPPING 1989). In the Tertiary, basaltic melts and fluid components (primarily $H_2O$ and $CO_2$) intruded this rock sequence causing in part extensive mineral alterations and fixation of large volumes of gas in the evaporites.

Still other possibilities for the formation and origin have been considered by various authors: for example, the thermal release of $CO_2$ from carbonates during contact metamorphism, the dissolution of carbonates by aqueous solutions with low pH values, and the fixation of $CO_2$ from the atmosphere (e.g., KÄSTNER 1968; FREYER 1973, 1978; FREYER & WAGENER 1975). However, it is seldom possible, if at all, to relate the portions of $CO_2$ in the gas-bearing evaporites to the aforementioned processes.

## 14.7.4 Hydrogen

The following three processes are considered for explaining the hydrogen enrichments in the gas-bearing evaporites (particularly in carnallitite):

1. During crystallization of carnallite, $MgCl_2$ can be replaced by small amounts of $FeCl_2$ (PRECHT 1879, 1880, 1905; BOEKE 1909, 1911; JOHNSEN 1909a, 1909b). $FeCl_2$ should break down part of the water of crystallization in carnallite into hydrogen and oxygen. Fe(II) is oxidized to Fe(III). Via intermediate stages $Fe_2O_3$ is supposed to form, which can be found as hematite in the carnallite crystals. According to ERDMANN (1910), however, experimental evidence of the course of this reaction under geological conditions cannot be provided.

D'ANS & FREUND (1954: 8) also discuss the oxidation of Fe(II) to Fe(III) due to steam at higher temperatures in the context of rinneite formation, during which hydrogen forms due to reduction of $H_2O$.

2. ERDMANN (1910) was the first to attribute the hydrogen in gas-bearing carnallitites to the decomposition of water under the influence of radioactivity (radiolysis). The natural radionuclides presumably find their way into the carnallite as daughter nuclides of the $^{238}U$ decay series (e.g., $^{226}Ra$) during the crystallization of salt minerals. Hydrogen is produced due to the effect of radioactivity on attached moisture, whereas oxygen is used in the oxidation of Fe(II) to Fe(III). ERDMANN (1910) explained the presence of nitrogen in the hydrogenous gas mixtures with the decomposition of $NH_3$, which also occurs in minor quantities in carnallite (BILTZ & MARCUS 1909; see also Chapter 14.7.5). ERDMANN (1910) regarded the occurrence of helium in the hydrogenous gas in the Staßfurt carnallitite of the Leopoldshall VI mine (Magdeburg-Halberstadt mining district) as proof of these formation processes.

The process of radiolysis is also being studied with reference to the underground disposal of high-level radioactive wastes in rock salt. It is known from these studies that, for example, hydrogen and oxygen form as reaction products due to the effect of radioactivity on aqueous NaCl and $MgCl_2$ solutions (e.g., JENKS 1972; JENKS et al. 1975; see also JOCKWER 1984, 1985; JOCKWER & GROSS 1985).

Formation waters, above all those related to petroleum and natural gas occurrences, contain, for example, the following amounts of natural radionuclides: 0.1 to > 20 µg U/1000 ml, 0.01 to > 1 ng Ra/1000 ml (BORN 1936: 43; see review in KREJCI-GRAF 1978: 15, 100 ff, 106). It is known that such formation waters or water accompanying petroleum penetrated into evaporites and, acting as metamorphic solutions, were the cause for salt mineral alterations (e.g., BORN 1934/1935, 1936; HERRMANN 1961a, 1961b). During this time, the gaseous products of radiolysis were able to be bound in the evaporites. In contrast, there is no clear evidence to support the assumption of ERDMANN (1910) that salt minerals already absorb radionuclides from seawater during their crystallization.

The $^{40}Ca$ and $^{87}Sr$ isotopes fixed in salt rocks form during the radioactive decay of the radionuclides $^{40}K$ and $^{87}Rb$. The strong reducing properties of these isotopes are probably the cause for the formation of hydrogen in evaporites in a reaction with water (SAVCHENKO 1958).

The occurrence of blue rock salt in gas-bearing evaporites could be a further criterium for the influence of radionuclide-bearing solutions on salt minerals (e.g., ERDMANN 1910; VALENTINER 1912; BORN 1934/1935, 1936, 1959; KIRCHHEIMER 1978). According to analyses of PANETH (in BORN 1934/1935: 79, 1936: 41) blue rock salt obviously contains more helium than colorless salt does (see also PANETH & PETERS 1928a, 1928b).

3. Hydrogen is also produced by microbiological processes. However, this hydrogen is in many cases used up in the reduction of nitrogen, sulfur, and oxygen compounds. Nevertheless, hydrogen is a widespread constituent of many natural gas occurrences (e.g., HUNT 1979: 153, 173). Since hydrogen is very mobile and easily reacts with other compounds, it is obviously consumed quickly in a geological

reservoir. Hence, the presence of hydrogen in a gas occurrence is a sign that there is constant rebonding of this element in the surrounding rock or diffusion from deeper strata (HUNT 1979: 173). In this context an observation of KNABE (1989: 360) is of interest: hydrogen in the Thüringen and Hessen potash salt seams nearly always occurs with elevated hydrocarbon contents.

Evaporites are undoubtedly able to fix hydrogen-rich gas mixtures in certain mineral associations for longer periods of time. So far gases have been found mainly in carnallitic rocks. This is probably related to the comparably easy mobilization and deformation of carnallite, bischofite, and tachhydrite. In spite of the concurrent iso-chemical and isophase recrystallizations (dynamic metamorphism), the more mobile and 'dry' phases like hydrogen are obviously fixed so firmly in the grain fabric of carnallite, bischofite, and tachhydrite that they are no longer able to be released by diffusion. Rock salt and sylvinite apparently do not have similar properties for fixing 'dry' hydrogenous gas mixtures. As described, for example, by SAVCHENKO (1958), the occurrence of hydrogenous gas mixtures in sylvinites from Solikamsk is apparently only possible with the simultaneous influence of gases and aqueous solutions (solution metamorphism) on the carnallitic rock. The sylvinite produced in this way is able to fix mineral-bounded gas.

## 14.7.5 Nitrogen

Nitrogen frequently dominates in gas mixtures fixed in the original rocks of the Zechstein evaporites (e.g., KNABE 1989; Tab. 9). Various processes are probably involved in the formation of nitrogen in gas-bearing evaporites as well. Nitrogen can form from organic matter via $NH_3$ and other reactions. Among other things, it is presumed that nitrogen can form due to reaction between $Fe_2O_3$ and nitrogenous organic constituents. The thermal decomposition of organic nitrogen compounds finely dispersed through shales supposedly leads to the production of organic nitrogen as well (e.g., HUNT 1979: 164).

The red bed rocks underlying the various evaporites are another source of nitrogen and gases containing high amounts of nitrogen. The Rotliegend rocks of the lower Permian (sandstones) underlying western European Upper Permian evaporites (e.g., the east Hannover gas province between Wustrow and Munster, Germany, see PHILIPP & REINICKE 1982) are an example.

Part of the nitrogen in gas mixtures enclosed in evaporites affected by magmatic processes is surely of inorganic origin, i.e., released during mantle degassing.

The question of whether a portion of atmospheric nitrogen can also be incorporated into evaporites during crystallization of minerals from seawater is still open (FREYER 1973, 1978; FREYER & WAGENER 1975). GROPP (1919: 35) attributes the occurrence of nearly pure nitrogen in evaporites to air which lost its oxygen during oxidation. It is assumed that the nitrogen occurrence in rocks underlying the Kupferschiefer of the Sangerhäuser and Mansfelder synclines are of similar origin (e.g., KURZE & GÖRING 1964).

In rare cases an indication of the source and formation of nitrogen may be obtained by determining the $^{15}N/^{14}N$ ratio.

BILTZ & MARCUS (1909) describe the nitrogen compounds ammonia and nitrate in the evaporites of the Staßfurt and Vienenburg potash salt mines.

## 14.7.6 Hydrogen sulfides

The hydrogen sulfide occurring in mineral-bound gas mixtures is the most toxic gas formed under natural conditions.

Hydrogen sulfide forms during the bacterial reduction of sulfates in seawater and formation waters. $H_2S$ participates in further reactions producing sulfur, metallic sulfides, and organic sulfur compounds. Hydrogen sulfide also forms from organic sulfur compounds and in the reaction between hydrocarbons and sulfate minerals and dissolved sulfates in formation waters. However, both processes are first initiated when temperatures of about 100°C are reached when sediments subside to greater depths. Under these conditions the chemical decomposition of organic sulfur contents in finely distributed organic material begins later than the formation of methane (e.g., HUNT 1979: 170 f). Such reactions are apparently the source of hydrogen sulfide in the hydrocarbon gases of evaporites. An example of this could be the isolated occurrence of hydrogen sulfide in nitrogen-methane mixtures in the Braunschweig-Lüneburg salt mine (Aller valley, Niedersachsen, Germany; EHRHARDT 1980).

The formation of hydrogen sulfide in gas mixtures which penetrated into the evaporites during magmatism apparently has to be evaluated in a different way. It obviously has to be assumed here that the hydrogen sulfide is related to the formation of basaltic melts and fluid phases in the upper mantle.

Isotope studies provide indications of the source of native sulfur and sulfur compounds in the solid, liquid, and gaseous constituents of evaporites (for recent studies see KNIPPING 1986, 1987a, 1989, 1991).

KNABE (1989) conducted extensive studies on the occurrence and composition of sulfurous gases in approximately 200 evaporite samples from the Thüringen and Hessen potash salt seams (Werre-Fulda mining district). These studies showed that the distribution of sulfurous gases varied much greater than that of other gases. The portion of sulfurous gases is often 0.2 - 10 ml/t rock. Such low contents cannot be detected by smell. KNABE (1989: 366) pointed out that dimethylsulfide $(CH_3)_2S$ and methylmercaptane $CH_3SH$ occur nearly as frequently as $H_2S$. Sulfurous gases were not detected in only 15 out of 143 samples (carnallitic rocks, sylvinite, rock salt; KNABE 1989: 366). The highest contents of sulfurous gases (i.e., 1200 - 1600 ml/t or 0.12 - 0.16 ml/100 g rock) were measured in secondary sylvinites, kieseritic Hartsalz (Flockensalz), and in banks of rock salt of K1Th and K1H. The high contents of sulfurous gases in sylvinite and rock salt nearly always occur with elevated hydrocarbon contents. The total volume of gas in kieseritic Hartsalz can consist of up to 90 % sulfurous gases, in which case $H_2S$ dominates. This is, however, not the rule. A clear correlation between the portions of $MgSO_4$ and $H_2S$ in the rock is not possible. This also applies to the correlation between $CaSO_4$ and $H_2S$. None the less, there is evidence that there is a relationship between evaporites containing calcium and magnesium sulfates and the occurrence of $H_2S$. Of all rock types studied, (primary) carnallitic rocks have the lowest contents of sulfurous gases. Carbonyl sulfide (COS) and carbon disulfide ($CS_2$) are rare in secondary sylvinite and rock salt (KNABE 1989: 366).

## 14.7.7 Noble gases

Helium and argon (possibly neon) are often detected in evaporites (for He see, e.g., STRUTT 1908; ERDMANN 1910; VALENTINER 1912; PANETH & PETERS 1928a, 1928b; HAHN 1932a, 1932b; KARLIK 1939; KOCZY 1939; GENTNER et al. 1954; for Ar see, e.g., initial studies by HARTECK & SUESS 1947, and later studies by ALDRICH & NIER 1948; SMITS & GENTNER 1950; GENTNER et al. 1953, 1954).

Gases from the gas fields east of Hannover (Germany) which originate from the upper Rotliegend consist predominantly of methane and nitrogen and contain 0.03 - 0.2 vol% helium (PHILIPP & REINICKE 1982).

Various authors have been able to demonstrate that seawater is obviously saturated and supersaturated with noble gases. Noble gas enrichments have also been detected in hydrothermal solutions. The individual noble gases in such solutions have the same ratios as in the air. Hydrothermal solutions also contain $^4$He and $^{40}$Ar (see ALEXANDER JR. 1974: 2-I-5). Both isotopes are formed during the decay of radionuclides of the natural series $^{238}$U, $^{235}$U, and $^{232}$Th as well as $^{40}$K. There are many indications that the helium and argon are of radiogenic origin above all in the hydrogen-dominated gas mixtures (see Chapter 14.7.4 and, for example, ERDMANN 1910; BORN 1934/1935, 1936, 1959).

## 14.7.8 Other components

In addition to the major components methane and nitrogen, small amounts of mercury, i.e., 600-4000 µg Hg/m$^3$ gas (PHILIPP & REINICKE 1982), have also been detected rarely in the gases originating from the upper Rotliegend of the east Hannover gas province. Reliable data on the source of the Hg cannot yet be presented (carbon from the Upper Carboniferous?, Rotliegend volcanics?). Mercury has not yet been detected in gas from evaporites - probably because these gases were never analyzed for Hg.

## 14.8   What is the origin of the gases in evaporites?

Where do the gases found in evaporite deposits come from? Basically, the two following possibilities are conceivable:
1. The gases formed from substances within the evaporite itself and were stored there. However, the gases in an evaporite sequence may be able to migrate: the present-day reservoir rock for the gases is not necessarily the source rock.
2. Gases and solutions from outside (i.e., the surrounding and overlying rock) penetrated into the evaporite. See Chapter 14.7 regarding both possibilities.

The question of the mobility of gases and solutions has become relevant in connection with the underground disposal of toxic anthropogenic wastes in evaporites. It has been (and still occasionally is) maintained that evaporites are absolutely impermeable to gases and solutions (e.g., RICHTER-BERNBURG 1979: 191). Such statements are simply not supported by the results of studies on evaporites. There are documented examples of the migration of gases and solutions through evaporites. In fact, it has been proven in practically all previously mined evaporites in the world that gases and solutions can spread and migrate through all types of evaporites under certain geological conditions. This is to say that evaporites basically do not behave differently with respect to the migration of gases and solutions than other magmatic, sedimentary, and metamorphic rocks (see also Chapters 14.7 and 14.10).

A widespread erroneous assumption is that gases and solutions cannot migrate through rock salt due to its low porosity and permeability. However, the mobility of gases and solutions in all rocks is not dependent solely on porosity and permeability, but also upon paths in the form of fractures such as cracks, fissures, etc. Rock salt is one of the tightest rocks with regard to porosity and permeability alone (see also the investigations of FÖRSTER 1974, 1985). In the case of fractures, cracks, and fissures, rock salt and other evaporites can locally and temporally become permeable to solutions and gases. Hence, these paths are obviously of greater significance to evaporites than porosity and permeability.

In this context the differing deformation properties of anhydrite and rock salt under high pressure are of interest. The load of vertical, interbedded rock salt and thin layers of anhydrite can lead to the formation of microcracks along the strings of anhydrite (e.g., FÖRSTER 1985: 64). The thinly bedded anhydrite forms zones of weakness along which gases and solutions can migrate under certain conditions. There is obviously less a chance of this in pure rock salt.

According to BAAR (1977: 90) evaporites are impermeable to gases and solutions at depths greater than 300 m. This assumption is based, however, on selected examples which are not representative of all known studies on evaporites. It has been found again and again that gases and solutions are released from evaporites at depths greater than 300 m and even over 1000 m. Results of drillings into the Gorleben salt dome provide current examples (e.g., GRÜBLER & REPPERT 1983: 48ff; HERRMANN 1983b: 39, 1984a: 442).

Releases of methane at depths of 400-460 m are also described from the Belle Isle mine (Louisiana, USA; e.g., IANNACCHIONE et al. 1984). The fact that the present-day reservoir rock for gases and solutions is usually not the site of origin is proof that mobile components migrate within evaporites under certain geological conditions.

The prerequisites for the mobilization and migration of gases and solutions from rocks underlying the evaporites or the surrounding rocks into an evaporite body or within the evaporites have been met when stresses have built up in the evaporite producing cracks, fissures, gaps, and other paths. It is interesting to note in this context that residual stresses have obviously been preserved over geological time in the tectonized evaporites of the Gulf Coast (USA; HARDY JR. & MANGOLDS 1980). Similar observations, however, have not been made in the salt domes of northern Germany.

Evaporites are stressed and deformed whenever halokinetic and/or tectonic processes affect the evaporite or pressures build up in the evaporite due to magmatism. Examples of the former cases are found in the salt domes and domes in central and northern Germany as well as in the horizontal salt beds of the southern Harz mining district. In the latter case the evaporites of Zechstein 1 in the Werra-Fulda mining district are excellent examples for study.

The formation of fractures resulting from the cooling of salt rock during the ice age (cryogenic fractures) is also a possibility to be considered regarding salt domes in northern Germany (BAUER 1991).

Fractures are found in both horizontal and steeply inclined evaporite beds (see references in WEIß 1980; see also BAUER 1991). Although there is evidence that fractures are more frequent in horizontal salt beds than in steeply inclined salt beds (salt domes), the following fact must be considered for proper understanding.

Due to the rather plastic properties of chloride rocks, paths formed by fracturing only remain open for a limited time - on a geological time scale. In most cases fractures reheal - becoming impermeable again - so that their former existence is hardly, if at all, detectable. Healed fractures in evaporites are usually evident especially when the former voids have been filled with secondary minerals or when small displacements in prominent layers of anhydrite, kieserite, or argillaceous material are present within the evaporite sequence.

Fractures in anhydrite and argillaceous rocks are more easily recognized. Healed fractures in these rocks can be detected after long periods of geological time even when the original void has long been refilled with secondary minerals such as halite, carnallite, sylvite, etc. Chloride rocks bordering on intensely fractured anhydrite or argillaceous rocks may also have been stressed. The fractures which formed in the chloride rocks under peak stresses are no longer visible.

Fractures are not the only paths for gases and solutions in evaporites. The mineral reactions accompanying dissolution and crystallization also allow extensive material transport, especially in K-Mg-bearing rocks and potash salt seams (i.e., solution metamorphism, see BRAITSCH 1962: 206f, 1971: 266).

Microcracks, traversing salt crystals in various directions, apparently aid the migration of gases in evaporites. However, observations of such microcracks are to be interpreted prudently. For example, the fine cracks forming a mosaic of 2- to 4-mm-large fragments which were observed in a rock salt sample of Devonian age from a depth of 2173 m (Issatschki drilling, Dnjepr-Donez depression, USSR) by ANTONOW et al. (1958) can have formed due to relief of pressure or drilling action.

With regard to the final disposal of radioactive substances in rock salt KRAUSE (1983) conducted experiments on the migration of natural and artificial solutions in the Avery Island salt mine near New Iberia salt dome (Louisiana, USA). Unfortunately, no data on the composition of the solutions were given, but they were probably saturated NaCl and/or $MgCl_2$ solutions. He found that after a year of heating the rock salt, the solutions migrated back toward the borehole during the cooling period, after the heating equipment was turned off . The paths in the rock salt taken by the solutions were probably microcracks. LIEDTKE & KOPIETZ (1983) also observed the formation of paths in rock salt after a temperature gradient had been created. In the studies of KRAUSE (1983) the permeability in the rock salt increased 4-fold over the cooling period. The average migration velocity measured was 0.57 cm per year in a part of rock salt, in which the temperature only varied about 0.5 °C

These findings are confirmed by migration tests conducted in the inactive Asse salt mine (ROTHFUCHS et al. 1985). The release of water (i.e., the small amounts of moisture in rock salt) was investigated while heating Staßfurt rock salt (Na2) to about 210 °C around a borehole in the presence of a $^{60}Co$ source. The heating equipment was switched off after about 850 days. Paths formed near the heat source as it cooled. Within a few days ten times more solution ('water') exited through these paths than did during the 2-year-long test period.

The texture and structure of a salt deposit are undoubtedly the decisive influence on the diffusion of gas (ANTONOW et al. 1958). KNIPPING (1984) and KNIPPING & HERRMANN (1985) showed using a basalt-carnallitite contact zone (Thüringen potash salt seam, Werra-Fulda mining district) that the mobile elements ($H_2O$, $CO_2$) transported by the basaltic melt not only moved along fractures in the evaporites but also along the grain boundaries of minerals and microcracks. The solutions migrated locally as a 'solution front' (figuratively speaking) through the mechanically stressed evaporites. When the volumes of solutions and gas are relatively small, it can be seen how the mobile front finally stagnates. WALTHER & ORVILLE (1982: 257) give a width of 0.1-10 µm for cracks along which mobile phases can still migrate in silicate (pelitic) rocks.

The paths for silicate melts, solutions, and gases which develop in the course of magmatism can be traced vertically and horizontally up to several hundred meters in the evaporites and overlying rock. In the Hattorf mine (Werra-Fulda mining district), for example, there are basalt dikes at a depth of about 750 m in the Hessen potash salt seam, which intruded the entire Werra sequence and overlying Buntsandstein up to the surface (e.g., KNIPPING 1989).

Not only basalt dikes but also the paths for gases and solutions can be traced over such distances. For example, gases penetrated 200 m and solutions 100 m into the evaporites of the salt domes of Louisiana (USA) from the site of basalt intrusion (KUPFER 1980: 132). This means that in evaporites gases can move over even greater distances in certain paths and under certain geological conditions.

When gases migrate through rocks over great distances, the gas mixtures may be fractionated, as in column chromatography (e.g., HUNT 1979: 51). Such processes may also have to be taken into consideration to correctly interpret enrichments of certain components in a gas mixture.

Gas in evaporites can accumulate in two ways (e.g., Fig. 17):

1. »Dry« gas mixtures from the surrounding or underlying rocks penetrate into the evaporite and are stored in cracks, fissures, and other fractures over geological periods of time. Some of the $CO_2$-bearing gases in the evaporites of the Werra-Fulda mining district are an example of this.

   »Dry« gas mixtures can also come into contact with evaporites composed predominantly of mobile and easily deformable minerals like carnallite, bischofite, or tachhydrite. When solid and gaseous components mix in the course of halokinesis or tectonism, the gases can become fixed in minerals. Hydrogen-rich gas mixtures may be stored in this way (Chapter 14.7.4). Examples are carnallite rocks in Solikamsk and Beresniki (former USSR, Lower Permian), near Staßfurt (Germany, Upper Permian), and in northeastern Thailand (Cretaceous).

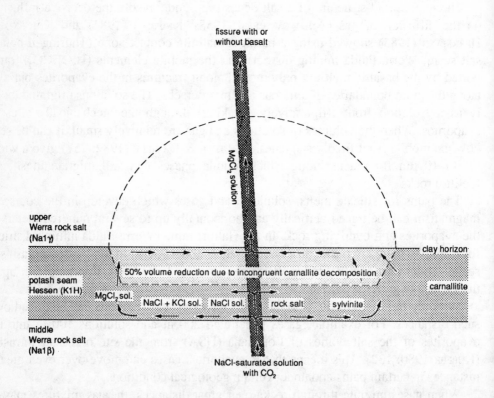

**Fig. 17** Possibilities for penetration and distribution of gases and solution in bedded evaporites in the Werra-Fulda mining district (schematic representation). The process was initiated by the intrusion of basaltic melts and fluid phases into the evaporites 13 - 25 million years ago. The semicircle delimits the zone of bearing pressure (after BAAR 1958: 145).

2. Gas mixtures together with aqueous solutions penetrate into the evaporites. When the solutions are not in equilibrium with the components of the surrounding rocks, existing minerals can be dissolved or decomposed, leading to the crystallization of new minerals. In this way part of the gas mixtures (including hydrogen-rich mixtures) are fixed in minerals, i.e., in the mineral grain itself or along its boundaries. Clear evidence of the originally collective occurrence of solutions and gases is found in recrystallizations in gas-bearing evaporites and in the coexistence of gases and solutions in microscopic inclusions. The formation of gas-bearing evaporites such as carnallitite, sylvinite, Hartsalz, and rock salt was in many cases only possible due to the common occurrence of both gases and solutions.

Gases occurring together with solutions probably constitute a considerable source of energy for solution migration, as can be inferred from measurements of the time dependence of solutions emanating from evaporites. The law of Boyle-Mariotte applies here whereby time dependence is seen as measure of the decrease in pressure. This is due to the expansion of gases (e.g., BRAITSCH 1962: 207, 1971: 267).

In nearly all cases, gas-bearing evaporites which formed from reactions with aqueous solutions contain only a portion of the original gases today. The following questions remain open: did the solutions participating in the reactions and the residual gases migrated into temporarily open paths and into overlying strata following the pressure gradient, or were they also able to migrate into the surrounding rock, now forming small solution and gas inclusions (HERRMANN 1979: 1084ff, 1991c; see also Chapter 13).

## 14.9   Release of gases during mining operations

The evaporite rocks of most known salt deposits to date contain more or less large volumes of gas. The amounts of gas contained in a single evaporite deposit can vary considerably even from one part of a mine to another; i.e., gas is inhomogeneously distributed through the deposit. Mining can become problematic particularly when large volumes of gas are stored in the evaporite being mined, or when gases are released into mine rooms from the over- or underlying rocks.

The release of gas from the rock into mine rooms can cause problems during salt mining especially when the gases are inflammable, like methane, other hydrocarbons, or, more rarely, hydrogen. In the past, gas mixtures containing methane or other hydrocarbons have penetrated into mines leading again and again to serious and fatal accidents, particularly in the Staßfurt sequence in the potash salt mines of the southern Harz mining district (e.g., BAAR 1952, 1954; LIEBSCHER 1952; STOLLE 1953, 1954; GIMM & ECKART 1968; MIETH et al. 1989).

Gas is also known to occur in salt mines - both in horizontal beds and in salt domes - in the eastern Alps, Poland, Russia, France, Spain, and North America (e.g.,

GIMM & ECKART 1968). In nearly all cases the gases are released into the mine through drill holes, fractured rocks, and other paths.

The difficulties in mining gas-bearing evaporites can also be illustrated with the potash salts in the Werra-Fulda mining district. The contents of $CO_2$-bearing gas mixtures in some Hartsalz and carnallitic rocks in the Thüringen and Hessen potash salt seams are so high (see Chapter 14.5) that there are still recurrent gas outbursts during mining today. The evaporites in the Menzengraben mine of the Heiligenroda potash works in the eastern Werra region contain particularly large volumes of gas. The greatest gas explosion known to date in potash salt mining occurred on 7 July 1953: Several 100 000 m³ of gas were released and about 100 000 metric tons of salt were displaced at the level of the Thüringen potash salt seam (e.g., JUNGHANS 1955; MÜLLER 1958; DUCHROW 1959; GIESEL et al. 1989).

## 14.10 Gas and isolated petroleum occurrences in the Zechstein evaporites of northern Germany

The salt domes and domes of northern Germany provide examples of the spatial relationships between evaporites and accumulations of gaseous and liquid hydrocarbons. Here, natural gas and petroleum occur frequently in the salt diapirs and domes consisting primarily of Zechstein evaporites. Since mining of the evaporites in northern Germany began, gas and isolated petroleum occurrences have been discovered in various evaporite horizons. The spatial relationships between evaporites and hydrocarbons are also known - on a greater scale - from the salt diapir region on the Gulf Coast of southern Texas and Louisiana (USA), for example.

The gas and oil occurrences of northern Germany have been described by FIEGE (1934, with 300 references).

In the following, some of the available data will be used to give a general description of the gas and subordinate petroleum occurrences in evaporites of northern Germany. The data may not be complete.

### Salt domes in which gases and petroleum have been found

Fig. 18 shows the geographical distribution of north German evaporites from which gas releases have been observed. In addition to gas, petroleum resp. condensates have been observed in evaporites mentioned under 4, 8, 15 and 16. In Fig. 18, the fact that the majority of localities (arrows) with gas releases are in the area of Hannover is not due to the geology there, but rather to the frequency of salt mining in this part of Niedersachsen.

References: NETTEKOVEN & GEINITZ (1905), GROPP (1919), FIEGE (1934), DEECKE (1949), ERLER (1957) in LÖFFLER & SCHULZE (1962), LANG (1973), MALZAHN (1973), EHRHARDT (1980), BAUER (1990). Name of the Evaporite bodies after JARITZ (1973).

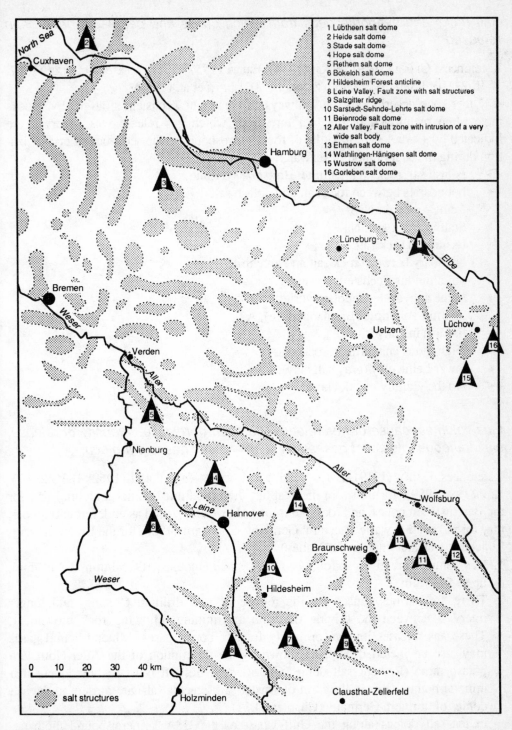

**Fig. 18** Evaporite bodies in northern Germany (after JARITZ 1973) in which gases, petroleum resp. condensates occur; evaporite bodies in the eastern part of Germany are not completely marked.

*Salt rocks and evaporite beds from which gases and petroleum have been released*

References: GROPP (1919), FIEGE (1934), ERLER (1957) in LÖFFLER & SCHULZE (1962: 183f, 231ff), LANG (1973), MALZAHN (1973), MIETH et al. (1989).

Older rock descriptions don't always fit clearly in Zechstein stratigraphy. Rocks most often mentioned in the literature from which gas release was observed are indicated by ++ or +++. The order of rocks corresponds to the sequence of crystallization during the evaporation of seawater.

++ potash salt rocks (carnallitite, Hartsalz, sylvinite)
++ boundaries between rock salt and potash salts
 Leine rock salt, Na3
 Staßfurt rock salt, Na2
++ fissures in rock salt
 boundary between rock salt and anhydrite
 »bituminous Zechstein gypsum«
+++ Leine anhydrite (Hauptanhydrite), A3
 Staßfurt anhydrite (Basalanhydrite), A2
++ fissures in anhydrite
 bituminous anhydrite in rock salt
++ lower Leine clay (gray salt clay), T3
++ »in the vicinity of salt clay«

*Gas volumes, cause of gas release, duration of release, intensity of release, simultaneous release of gases-salt solutions and gases-petroleum*

References: GROPP (1919), ERLER (1957) in LÖFFLER & SCHULZE (1962: 183f, 231ff), HEMMANN (1989), for mines in the Staßfurt area (Staßfurt-Egelner anticline) and for the Bartensleben potash and rock salt mine (Aller valley), used as underground repository for radioactive waste in central Germany, EHRHARDT (1980) for the Braunschweig-Lüneburg rock salt mine (Aller Valley).

1. Gas volumes are given as follows: traces, small amounts, 10-15 l/min, 2-3 m³/min, up to 5 m³/min.
2. There are various causes for the release of gas: drilling, blasting, and longer intervals between the driving of drifts and mining following rock movement. These gas releases involve primarily fracture-bound gases. Mineral-bound gases may also be included here, as shown by the description of the Aller-Nordstern potash mine (Rethem salt dome; GROPP 1919: 36f). It is to be noted here that mineral-bound gases in the form of knister salt (crackle salt) are present in the salt domes of northern Germany (Chapter 14.11).

In the salt domes along the Gulf Coast of the USA, mineral-bound methane-bearing gases were released in the same manner as $CO_2$-bearing mineral-bound gas mixtures in the Werra-Fulda mining district. For example, in the Belle Isle Mine (Louisiana) voids produced by gas outbursts had diameters of 1-10 m and heights of 1 to > 22 m (IANNACCHIONE et al. 1984). This corresponds to about 0.8-1700 m³ or 1.8-3700 metric tons of displaced rock salt.

3. Gas releases can last for minutes, hours, days, months, and years. In the Bartensleben mine a gas release observed 1942 in anhydrite was still active in 1950.

4. Gas releases can be described in terms of volume, type of storage (in fractures or mineral-bound), and composition as follows: weak releases, weak and strong drafts, releases accompanied by loud noise, gases under high pressure, gas fires, relatively seldom firedamp explosions.

5. BAUMERT (1928: 69) noted that gas releases often lack solutions. In contrast, releases of solutions from evaporites are frequently accompanied by gases, especially in the initial stages. The latter are usually mixtures consisting of $CH_4$, $CO_2$, $N_2$, and $H_2S$ (BAUMERT 1928: 69; after GROPP 1919). GROPP (1919: 39) described such a case from 1914 at the Teutonia potash mine near Wustrow (Wustrow salt dome, Wendland, southwestern Gorleben salt dome see Fig. 18) as follows:

»During investigations of the north-south-striking carnallitite potash seam two voids filled with secondary flaky carnallitite were uncovered. There was a distinct smell in the voids similar to gasoline. Gases and brines were released accompanied by a crackling noise. Analysis yielded a gas composition of 9.13 % $CH_4$, 16.3 % O, and 74.37 % N. The presence of firedamp in the voids could still be detected days later with a safety lamp even after extensive ventilation. After a portion of concentrated seawater in the void had crystallized, the gases apparently migrated into fissures and cracks in the void.«

6. GROPP (1919) reported several simultaneous releases of gas and petroleum from evaporites between 1907 and 1917. In addition to hydrocarbon-bearing gases, small amounts of petroleum were also observed at several places on the mining rooms of the Teutonia potash mine (Wustrow salt dome, Wendland). Gas mixtures and petroleum were also observed in the Hope salt dome and in the evaporites of the Leine Valley.

7. The simultaneous release of solutions and gases was also observed in the salt domes of northern Germany. In this context it is to be mentioned that fractures in chloride rocks extending down to 600-700 m below the surface have been observed in the extensively investigated salt domes of Niedersachsen (BAUER 1991). Fractures in brittle rocks (anhydrite, argillaceous rocks) were detected down to even greater depths. Only such fractures (fissures, voids) are important as paths for salt solutions, gases, and condensates. In comparison, the low porosities and permeabilities in rock salt have, of course, hardly any significance.

## Composition of the gases

References: GROPP (1919), FIEGE (1934), ERLER (1957) in LÖFFLER & SCHULZE (1962: 231-334), MALZAHN (1973), EHRHARDT (1980), KNABE (1989).

Hydrocarbons - predominantly methane and nitrogen - have often been reported in gases occurring in the salt domes of northern Germany. According to GROPP (1919), the gas mixtures released from the carnallitite and adjacent rock salt carnallitite in the Wustrow salt dome (Teutonia potash mine, Wendland) contained 3 - 14 % methane and 16 - 19 % oxygen, the rest making up with nitrogen.

In 1936 oil impregnations and gas were discovered in the sandstones on the southern flank of the Hope salt dome. The gas had the following composition: 80.8 vol% $CH_4$, 0.8 vol% $C_2H_2$, 1.2 vol% CO, 0.2 vol% $CO_2$, 11.6 vol% $N_2$ (MALZAHN 1973: 65). The gas tubes (holes after gas outburst) in the area of the so called oil cracks in the Hope salt mine supposedly contained a high amount of nitrogen (fossil air, MAYRHOFER 1973: 49).

According to ERLER (1957) in LÖFFLER & SCHULZE (1962: 230) gases in the Bartensleben mine (Aller Valley) consist predominantly of mixtures with 20-30 % hydrocarbons (mostly methane, sometimes with ethane, $C_2H_6$), 60-80 % nitrogen, several percent hydrogen and oxygen, and rarely carbon dioxide. An exception is the gas mixture released by a carnallitite composed of 25 % hydrogen, 63 % nitrogen, 12 % oxygen, and only 0.2 % methane. A gas mixture of 65 % methane and 35 % nitrogen (released from the boundary between rock salt and anhydrite) was regarded as another extremely rare case. In one part of the Bartensleben mine the gas mixture contained 70-80 % nitrogen and only 2-25 % methane, while oxygen reached 12-19 %. In addition, 0.2-0.3 % $CO_2$, no hydrogen, and isolated occurrences of hydrogen sulfide and noble gases (e.g., helium) were determined in the gas mixtures.

The gases in the Braunschweig-Lüneburg rock salt mine (also Aller valley) had similar compositions. Besides dominant nitrogen, they contained 7-45 % methane and, in rare cases, up to 0.085 % hydrogen sulfide (EHRHARDT 1980).

Based on available data on gases in the individual evaporites gases of very differing composition can presumably be encountered in different areas of the salt mine. Nevertheless, for the salt domes of northern Germany it should be possible to generalize that nitrogen and hydrocarbons are usually the dominant constituents in the salt-bound gases.

## Origin of the gases in the salt domes of northern Germany

The spatial relationships between hydrocarbons and evaporites can be traced back to the relationships between accumulations of natural gas and petroleum and salt structures (LOTZE 1957). Not all of the questions connected with this subject will be dealt with in this chapter. However, three processes by which gases, petroleum, or condensates can obviously accumulate in evaporites must be discussed (see also Chapter 14.8).

1. During the ascent of a salt body the beds of the surrounding rock are dragged upward along the sides of the salt dome. In their subsequent position, the ends of the surrounding rock strata butt up against the evaporite body. Such structures can serve as traps for hydrocarbons since salt is often a barrier for gas and petroleum, preventing their further extensive migration. None the less, hydrocarbons have been observed to penetrate into salt domes primarily through local paths along the margins. An example is the petroleum (release of the so-called »Öl-Knacks«; e.g., LANG 1973: 67) and gas (HEIDORN 1926; MALZAHN 1973) encountered in the overhangs of the Hope salt dome. These penetrations of hydrocarbons from the adjacent rocks into the salt may be related to the intense deformation of the rock during ascent of the salt. There is much evidence that the transport of gas and oil is also aided by salt solutions. In fact, gas-bearing evaporites frequently show clear evidence of altered and neogenic minerals. Thus, MAYRHOFER (1973: 49) believes that oil formation water may have penetrated into the evaporite during ascent of the salt dome.

   The possibility that gases penetrated into the evaporite is also to be considered when gaseous compounds occur as deposits in rocks underlying the evaporite sequences (e.g., eastern Hannover gas province; PHILIPP & REINICKE 1982). However, such occurrences of gases below evaporite sequences do not necessarily have to be genetically related to the gases in evaporites.

   After an ascending salt has slowed considerably or come to rest, there no longer have to be any intact paths between the salt dome and adjacent rock for the gaseous and liquid hydrocarbons or salt solutions. Likewise, during renewed deformation of the salt, old paths between the salt dome and adjacent rock can be reactivated.

2. Obviously, organic substances can accumulate directly in evaporites during formation of a saline sequence. In such cases gaseous and liquid hydrocarbons are present in evaporites without the penetration of gases or petroleum into the salt dome from outside.

   The present-day reservoir rock for hydrocarbons within a salt dome are often not the original site of deposition of the organic substances. In the course of deformation during ascent of the salt dome, gaseous and liquid hydrocarbons - probably together with salt solutions - were able to disperse vertically and horizontally through the evaporite along the pressure gradient. LOTZE (1957: 383) points out that the migration of petroleum with densities between 0.7-0.97 $g/cm^3$ is aided by the presence of concentrated salt solutions. Hydrocarbons and salt solutions penetrate into strata through the more favorable paths. This applies particularly to beds of anhydrite in salt domes.

3. In the salt domes along the Gulf Coast of the USA several individual bodies joined to form one salt dome during the ascent of the evaporite. In this case nonevaporites such as sandstone, shale, and clay were also enclosed in the evaporite. These rocks frequently contain organic substances and solutions. It is presumed that the origin

of the gases (methane) and solutions in salt domes of the Gulf Coast can be explained in this way. The gases and solutions are found in zones of anomalies in the rock salt (e.g., IANNACCHIONE et al. 1984).

The individual case has to be examined as to whether it falls into any of the three categories mentioned regarding gas and petroleum. Several causes may be conceivable. Isotope studies ($\delta^{13}C$, $\delta D$) are helpful when investigating the formation and origin of gas mixtures and condensates in evaporites (Chapter 14.11). Few isotope data on gases from the salt domes of northern Germany have been published (e.g. GERLING et al. 1991). Therefore, the possibility for reliable comparison of the hydrocarbons from within and around the salt domes are currently limited.

## Do gases pose a threat to salt mining in Niedersachsen (Germany)?

The assessment of GIMM & ECKART (1968: 553) that there is no great danger of gas in the salt mines around Hannover and in Niedersachsen is still valid today. However, SPACKELER (1957) should be quoted here: »Unfortunately, much information on this subject is currently being lost due to mine closures particularly in Niedersachsen. Four cases are known due to violent explosions, i.e., Frisch-Glück (1904, 16 fatalities) in anhydrite, Desdemona (1906) in carnallite, Aller-Nordstern (1911) in carnallite, and Adolfsglück (1912) while the shaft was being sunk in rock salt (all Hannover district).«

The aforementioned salt mines were situated in the following evaporite bodies indicated in Fig. 19: Adolfsglück = Hope salt dome (4), Aller-Nordstern = Rethem salt dome (5), Frisch-Glück and Desdemona = Leine Valley (8).

The explosion in Adolfsglück (Hope salt dome) occurred at 517 m depth as a shaft was being sunk. The gases were released from a mixture of rock salt, anhydrite, and clay and were ignited by carbide lamps. Five people died in the accident. As a result, an attempt was made to bypass this gas-bearing zone by drifting away from this shaft. GROPP (1919: 40) describes the experience gained as follows: 'In aftermath of the explosion at 517 m in the shaft being sunk in December 1912, killing five people, work on the shaft was discontinued, and a floor was laid at 500 m where drifting was begun. In the course of this work gas-rich oil was often encountered in horizontal drillings and drifts. There was a very considerable outburst on 25 August 1913 from a horizontal drilling toward the northeast with a length of about 135 m. Approximately 140 000 kg of petroleum flowed out of this borehole. The oil was under such a high pressure that the entire drilling pipe was blown out of the hole into the drift, part of it being bent and coiled in the process. The oil flowed into the shaft sump as gas filled the mine. The crew was able to make it to safety, with two exceptions. These two men fell unconscious in the shaft, but were able to be saved and revived in time by the rescue team. It was not possible to determine from which rock the oil originated because the drill core was not recovered. It is assumed that the oil was released from a fissure in anhydrite or at one of its boundaries. The gas was not analyzed, but it is assumed to be hydrocarbons.

Today as well, accumulations of gas are released into mines during drilling and underground mining of salt in Niedersachsen (e.g., EHRHARD, 1980).

## 14.11 Gases in the Gorleben salt dome

Gases and liquid condensates were encountered in shaft drillings Go 5001 and Go 5002 in the Leine rock salt (Na3) in the center of the Gorleben salt dome (e.g., AKSTINAT 1983; GRÜBLER 1983, 1984a, 1984b; GRÜBLER & REPPERT 1983; PTB aktuell 1983, no.10; SCHOELL 1983). In addition, core material from deep drillings Go 1002, Go 1003, and Go 1004 on the flanks of the salt dome was analyzed for gaseous hydrocarbons (SCHMITT 1987; GERLING et al. 1991).

### *Data on shaft drilling Go 5001*

Two gas and condensate occurrences were registered, both in the Leine rock salt (GRÜBLER 1984: 167; GRÜBLER & REPPERT 1983: 48ff). 1. between 864.5 m (870.3 m) and 872.1 m (Na3$\delta$, $\varepsilon$), and 2. between 940.0 m (960.2 m) and 967.6 m (Na3$\gamma$, orange salt).

The gas released as a mixture of hydrocarbons containing methane ($CH_4$), ethane ($C_6H_6$), propane ($C_3H_8$), and traces of butane ($C_4H_{10}$).

The condensate entering the borehole at 864.5 m was evidenced by a pulsating increase and decrease in the volume of drilling mud. There was the smell of diesel and bubbling (PTB aktuell 1983).

Narrow fractures were detected at depths of 871.35 m, 871.40 m, and 966.55 m in shaft predrilling Go 5001 (GRÜBLER & REPPERT 1983: 49). However, it is not known at the present whether these fractures already existed in the evaporites or were produced by drilling.

Around 4 m³ of condensate flowed into the drill pipe between 855 m and 872 m within several days. Between 940.0 m and 967.6 m the volume of condensate entering the pipe over several days was only about 0.3 m³. However, the condensate at greater depth had already been released before measurements were taken (GRÜBLER 1984a: 167ff).

This condensate was a mixture of lighter and heavier hydrocarbons. It entered the borehole under overburden pressure as a liquid and passed in part into the gas phase before reaching the surface due to the release of pressure, indicated by the formation of bubbles in the drilling mud. The liquid components of the condensate remained in the drilling mud (PTB AKTUELL 1983).

## Data on shaft drilling Go 5002

The first hydrocarbon releases were detected four meters below the salt wash surface of the salt deposit (254.8 m) and continued down to a depth of 315 m. There were no indications of gas between 315 m and 328 m. Gas releases then resumed, again increasing in volume down to 355 m and subsequently decreasing down to 382 m. There were no indications of gas between 382 m and 426 m. Continuously varying gas releases were registered down to the final depth of 965.0 m (GRÜBLER 1984a: 174).

The gases were released from the Leine rock salt (Na3, orange salt). In addition to methane, the gases also contained small amounts of ethane, propane, and butane.

The releases of condensate and gas were small. Two days after the borehole was emptied down to a depth of 814 m, a constant release rate of < 100 l per hour was measured (GRÜBLER 1984a: 177).

## Occurrences of crackle salt (knister salt)

The gases in the Gorleben salt dome are not only found in cracks and fissures, they are also trapped in minerals. For example, crackle salt was encountered in shaft drilling Go 5001 at a depth of 966.55 m (Na3, orange salt). Here, the gas is fixed under pressure in the mineral grains and along grain boundaries (GRÜBLER & REPPERT 1983: 49).

In shaft drillings Go 5001 and Go 5002 no core material was lost due to releases of gas (GRÜBLER 1984a: 179). It is further noted that there was no rock salt ejection was not observed in either drilling Go 5001 or Go 5002 (GRÜBLER 1984b: 436f). A rock salt ejection is caused by the sudden release of mineral-bound gases. However, the fact that there was no rock salt ejection does not conflict with the occurrence of crackle salt. Although gases in cracks and fissures are released during drilling, this is rarely the case for mineral-bound gases. Mineral-bound gases are commonly released abruptly as a result of strong vibrations, above all from blasting (see also Chapter 14.10).

In the case of the shaft drillings in the Gorleben salt dome, it is presently not known whether the gases were contained mostly in cracks and fissures or in minerals.

## Potash salts as preferred gas-bearing horizons

A comparison of data on the gas releases in the salt domes of northern Germany with those of previous gas occurrences in the Gorleben salt dome is of interest here. In our opinion such a comparison can lead to considerations which have not yet been discussed.

It was pointed out in Chapter 14.10 that gases and condensates from external sources have been observed to penetrate primarily into the marginal zones of the salt dome (e.g., Hope salt dome). Four deep drillings were sunk in the margins of the

Gorleben salt dome. Solutions were released in all four boreholes (e.g., HERRMANN 1983b: 38, 1984a). In contrast to shaft drillings Go 5001 and Go 5002 in the center of the salt dome, however, no notable releases of gas or condensate were mentioned in the well logs.

It is pointed out here, however, that gases containing 4 - 7640 µg $CH_4$/kg salt rock (ppb) were able to be isolated from cores of exploratory drillings Go 1002, Go 1003, and Go 1004. There are no data on the total volume of gases released (SCHMITT 1987). Thus, the gas analyses on these three drill cores are not considered in the following discussion.

It is known that releases of solutions from evaporites are sometimes accompanied by gases. There are numerous examples of this in Niedersachsen, e.g., the Wustrow salt dome near the Gorleben salt dome (Fig. 18).

The following fact should be considered when interpreting the lack of notable gas and condensate releases in the four deep drillings in the Gorleben salt dome. Mineral-bound gases are fixed primarily during mineral reactions with aqueous solutions in the salt rocks. In Gorleben previous exploration has shown that the Staßfurt potash salt seam (K2) is composed predominantly of carnallitite. Hence, it can be presumed that at least these zones of the evaporite sequence have had no or very restricted contact with unsaturated aqueous solutions since their formation 250 million years ago. This means that there have also been no mineral reactions caused by solutions, and consequently, no mineral-bound gases in greater volumes were able to be fixed in the carnallitite of the Staßfurt potash salt.

GRÜBLER (1983: 37) brought a further argument. The hydrocarbon-bearing rocks of the upper Keuper, Lias, and Dogger do not border the flanks of the Gorleben salt dome. Hence, petroleum cannot migrate into the Gorleben salt dome, as in the case of the Hope salt dome.

The fact that large volumes of gases and petroleum are lacking in the previously investigated marginal zones of the Gorleben salt dome is remarkable mainly because the potash salts of the salt domes of Niedersachsen are often potential hosts for gases (Chapter 14.10).

According to older descriptions, however, the original composition of these potash salts has frequently been altered by mineral reactions involving solutions. Residual solutions and gases are then trapped in the evaporites (e.g., Wustrow salt dome, Chapter 14.10).

The high amounts of $MgCl_2$ in the solutions of the Gorleben salt dome can be attributed to reactions involving originally unsaturated aqueous solutions and the mineral carnallite (Herrmann 1984a). However, in Gorleben the following question is yet to be answered: From which parts of the salt dome did the solutions now stored in the Leine anhydrite (Hauptanhydrite) and Staßfurt rock salt actually obtain their composition?

## Composition of the gas mixtures in the shaft drillings

The gas samples collected from shaft predrillings Go 5001 and Go 5002 were in part mixed with air. There were also samples with nitrogen and carbon dioxide which could not be explained by air (AKSTINAT 1983). Various authors have studied the composition of the hydrocarbons and the fractionation of the carbon and hydrogen isotopes in the hydrocarbons (see »Data on shaft drillings Go 5001 and Go 5002« and the following section).

AKSTINAT (1983) obtained the following composition for gas samples from drilling Go 5002, which were obviously not contaminated with air: 55-88 vol% $N_2$, 10-36 vol% $CH_4$, 0.05-0.1 vol% $CO_2$, and higher hydrocarbons. In a mixture of hydrocarbons methane dominated with about 82 vol%, followed by 9 vol% ethane, 5 vol% propane, 2 vol% butane, and < 1.5 vol% other hydrocarbons (AKSTINAT 1983). The hydrocarbon-rich gases and $N_2$-$CO_2$ mixtures presumably have sources which vary greatly in composition (AKSTINAT 1983).

Isotope distribution has been a point of emphasis in previous studies on salt-bound gases from the Gorleben salt dome. Equally important are data on the quantitative composition of the gas mixtures (complete gas analyses). Of particular interest here are the portions of hydrocarbons, nitrogen, oxygen, hydrogen, carbon dioxide, carbon monoxide, and any hydrogen sulfide or noble gases (see Tab. 6). Such complete analyses form the basis not only for the interpretation of the occurrence and source of the gases but also for required safety measures.

In the course of surface exploration, representative gas samples for complete analysis were practically impossible to obtain. However, the release and analysis of the mineral-bound gases in knister salt might be possible.

## Isotope studies of gases and condensates from the shaft drillings

Using the $^{13}$C and D values for methane an attempt was made to obtain information on the source of gases in the Gorleben salt dome. The following values were yielded (SCHOELL 1983, see also GERLING et al. 1991):

| | $\delta^{13}C$ (‰) | $\delta D$ (‰) |
|---|---|---|
| Gaseous hydrocarbons | | |
| methane from the Na3γ | | |
| of the Gorleben salt dome | -45 | -170 |
| methane in gases from the | | |
| Rotliegend near the Gorleben | | |
| salt dome | -20 to -24 | -120 to -124 |
| Condensates | | |
| mean composition | $-26.2 \pm 0.2$ | $-65 \pm 2$ |

## Volume of the gas mixtures from the shaft predrillings

The gas volumes released mainly from the cracks and fissures in the shaft predrillings were able to be estimated (for details see 'Data on shaft predrillings Go 5001 and Go 5002'). Data on the volumes of mineral-bound gas in crackle salt would be equally important.

## Sources of the gases in the shaft drillings

There is much evidence indicating that the gases in Na3 of the Gorleben salt dome did not originate from pre-Zechstein rocks. However, a pre-Zechstein source for the condensates cannot be ruled out (SCHOELL 1983). It is presumed that the gases formed in situ after the condensates migrated into the evaporites. SCHOELL (1983) proposed two models for discussion: (1) the condensates migrated from a source at greater depth into their present reservoir, and (2) the condensates formed from substances in the immediate vicinity of their present occurrence.

Gerling et al. (1991) assume the Kupferschiefer being the source of the hydrocarbons in the evaporites of the Gorleben salt dome, with the exception of the potash seams.

## Which rocks in the Gorleben salt dome can act as reservoirs for gas?

In the Gorleben salt dome gases and condensates have only been found in the Leine rock salt (Na3γ, Na3δ, ε). As indicated by the findings for shaft drilling Go 5002, the gases and condensates in Na3γ (orange salt) are possibly distributed over great areas and depths (since beds dip steeply). The mineral-bound gases (crackle salt) are fixed predominantly during recrystallization in the presence of aqueous solutions, as mentioned above. Hence, the following questions should be considered regarding Na3γ: 1. Is the gas-bearing crackle salt particularly coarse grained and transparent and colorless, or brown (like condensate)? 2. How much Br does the rock salt contain? 3. Are there any indications of recrystallizations in Na3γ?

Based on observations from other salt domes in Niedersachsen and the present status of research, the following rock units must be classified as potential gas reservoirs, even if there is still no direct confirmation of this from deep drillings on the flanks of the dome (see also Chapter 14.10):

potash salt rocks
Leine rock salt, Na3γ, and other rock salt beds
fissures in rock salt
at the contact between rock salt and anhydrite
Leine anhydrite (Hauptanhydrite), A3
fissures in anhydrite
at the contact between anhydrite and salt clay
salt clay

## What gas volumes and compositions are to be expected in the Gorleben salt dome?

Hydrocarbon-bearing gas mixtures in particular are to be expected in the Gorleben salt dome? Based on previous experience in other salt domes the danger of firedamp explosions is not great, but cannot be ruled out completely.

Gas volumes are very difficult to predict. Based on experience in Niedersachsen release rates of several liters to cubic meters per minute are to be expected (e.g., EHRHARDT 1980). Releases can last from a few minutes up to several years in rare cases (Chapter 14.10). The threat posed by such releases is to be evaluated based on whether the gases are released in the initial stages of subsurface exploration (small voids) or in very advanced stages (large underground mined caverns, large voids). In the first case, extreme caution is to be exercised with respect to ventilation and gas protection.

The salt mines in Niedersachsen have all been in operation for about 100 years. In recent decades the relatively rare releases of gas have posed hardly any serious threat. In Gorleben the conditions with respect to gas surely do not differ fundamentally from those in other salt domes of Niedersachsen. Due to its toxicity particular caution is to be exercised with respect to local occurrences of hydrogen sulfide.

In the salt domes of Niedersachsen there have not yet been any voluminous gas releases, such as those known under the different conditions in the Werra-Fulda mining district. The releases of gas in the salt domes of Niedersachsen can be managed with mining techniques - such as sealing or concreting of boreholes, controlled degassing of gas-bearing zones, and bypassing of zones with high gas contents. In this way neither mining activities nor the mine itself are threatened (see, for example, EHRHARDT 1980, for the Braunschweig-Lüneburg salt mine, Aller valley). Based on current knowledge, a similar prognosis can be made for the Gorleben salt dome as well.

## 14.12 Adverse effects of salt-bound gases and radiolysis on the long-term safety of underground repositories

While heating rock salt samples (0.1 - 0.2 vol% $H_2O$) from the Asse salt mine to 300 °C JOCKWER (1984, 1985) and JOCKWER & GROSS (1985) recorded the release of the following gases (maximum values relative to NTP): 450 l $CO_2$/m³ rock salt, 6.2 l $H_2S$/m³ rock salt, 195 l HCl/m³ rock salt, and 156 l hydrocarbons/m³ rock salt. HCl is released primarily as a result of the thermal decomposition of certain salt minerals (e.g., SCHRADER et al. 1960; JOCKWER 1984: 19).

The fundamentals of radiolysis and its application to salt minerals will not be dealt with here; for information on this topic see the publications and reports of, for example, JENKS et al. (1975), JENKS & BOPP (1977), LEVY et al. (1981), VAN OPBROEK & DEN HARTOG (1985), and O. SCHULZE (1985). The theoretical assumptions of VAN

OPBROEK & DEN HARTOG on the extent of radiolytic effects of high-level wastes on a repository also cannot be discussed in this chapter. Only experimentally founded observations should be taken into consideration of the conceivable adverse effects on the long-term safety of underground repositories.

In connection with the subject of gases in evaporites, JOCKWER (1984, 1985) and JOCKWER & GROSS (1985) determined that also small amounts of HCl and $H_2S$ can form when natural rock salt with small quantities of $H_2O$ is exposed to gamma radiation. A relationship between the quantity of $H_2O$ in the rock salt sample and the gas components formed has not yet been established. They were also not able to recognize a relationship between the chloride content of the samples and the formation of $Cl_2$ and HCl. The mineralogy of natural rock salt obviously has a decisive influence on the formation of various gaseous components (JOCKWER & GROSS 1985). They point out that gases form due to the thermal and radiolytic decomposition of evaporites, above all in connection with high-level, high-heat-generating wastes. In this context both the corrosion of waste containers and release of radionuclides as well as the buildup of pressure in the boreholes with the stored wastes are to be considered (JOCKWER & GROSS 1985: 594). They recommend the detailed study of the parameters determining the formation of gas in the rock salt.

In this context the observation of JOCKWER (1984) that below 100 °C no HCl forms in rock salt due to either heating or gamma radiation is of interest. In contrast, gamma radiation does cause the formation of $H_2S$ in rock salt below 100 °C (JOCKWER 1984: 22). Past studies give no clue as to the possible quantities and conditions of gas formation during the radiolysis of evaporites and salt solutions. For example, no data is given for the volumes of gas formed during the solution migration tests in the former Asse salt mine (ROTHFUCHS et al. 1985: 131).

Based on the aforementioned investigations there is no doubt that underground repositories in rock salt for non-heat-generating, low-level radioactive wastes in salt domes are to be assessed differently from those for high-heat-generating, high-level radioactive wastes.

Which relationships exist between the gases trapped in the salt and those formed by radiolysis and the construction and long-term safety of an underground repository in a salt dome?

The problems caused by salt-bound gases during the geological and engineering investigation and construction of an underground repository are known and can be dealt with when previous experience is carefully considered and all safety precautions are taken.

Further considerations are necessary in connection with the long-term safety of underground repositories in salt domes. For instance, a differentiation must be made between nonradioactive, non-heat-generating and low-level radioactive wastes, on the one hand, and high-level radioactive, high-heat-generating substances, on the other.

At natural rock temperatures of < 100°C and with no radiation of the rock, the long-term safety of underground repositories in evaporites is considerably less influenced by 'dry' gases trapped in the salt than by aqueous saline solutions. This is to say that in the long-term the integrity of a repository for nonradioactive and low-level radioactive wastes is not endangered by the natural gases in the rock salt of evaporite deposits.

In contrast, the behavior of salt-bound and neogenic gases must be considered for the safe disposal of high-level radioactive and high-heat-generating wastes in rock salt. The laboratory experiments of JOCKWER (1984) and JOCKWER & GROSS (1985) are useful in this context.

However, there is the question of whether or not results of laboratory experiments are valid under natural conditions and over long period of time. Here it is pointed out again that field observations made in natural salt deposits also have to be included in the safety considerations. Above all, it is the time factor of the long-term natural processes that cannot normally be accounted for in laboratory experiments or model tests in natural bodies of rock.

An interesting object of study for the effects of high temperatures on rock salt are zones of contact between basalt dikes and the intruded evaporites of Zechstein 1 of the Werra-Fulda mining district. Here, 13 - 25 Ma ago basaltic melts with temperatures of around 1150 °C intruded into the rock salt with a temperature of about 50 °C. The resulting spatial and temporal temperature distribution was calculated for a »dry« basaltic melt by KNIPPING (1987a, 1989, 1991). For example, after intrusion of the melt the temperature at the contact between a 1-m-wide dike and the rock salt was 790 °C. Thirty days later, the temperature at the now solidified basalt was only about 400 °C and after one year it had decreased to less than 100 °C (KNIPPING 1987a, 1989, 1991).

Thermal decomposition of salt minerals caused by high temperatures would not be demonstrated. The mineral reactions and material transport which can be proven near the basalt dikes today can be attributed to fluid phases ($H_2O$, $CO_2$) brought by the basaltic magma. This is to say that the compositions of the basalts and evaporites were influenced above all by the aqueous solutions, and not by »dry« gas mixtures. Even the stresses of the high temperatures and concentrated salt solutions on the silicate compounds did not lead to an extensive decomposition of the entire silicate rock. Secondary minerals and recrystallizations are found primarily at the contacts between the basalt and rock salt and a few millimeters into the basalt.

These observations can be taken into consideration in the disposal of high-heat-generating wastes in the following way: The heat radiating from containers with high-level radioactive wastes may lead to thermal decomposition of salt minerals and to release of salt-bound gases near the heat source. However, this process should not cause any extensive mobilization of high-level radioactive substances and their migration into the overlying rock due to the lack of salt solutions.

In the comparison between the disposal of high-heat-generating substances in rock salt beds and the effect of basaltic melts on rock salt it is to be remembered, however, that heat from radioactive decay will be generated for a much longer time, compared with the cooling of a relatively small body of basalt in a salt deposit.

No observations have been made in natural salt deposits that are directly comparable with the radiolysis determined by JOCKWER (1984, 1985) and JOCKWER & GROSS (1985). In this context surely no conclusions can be drawn from the formation of blue rock salt. In any case the further study of the formation of small amounts of gases such as HCl due to radiolysis appears expedient (see JOCKWER 1984, 1985; JOCKWER & GROSS 1985).

## 14.13 Detection of gases and condensates in evaporites

Gases trapped in evaporites are released when driving shafts and during underground exploration. The prerequisite for effectively guarding against the threat posed by such releases is the prompt recognition of the problems and the localization of gas and condensate accumulations in the solid rock as well as the detection of certain components in the mine air.

The techniques used today for detecting gases in potash and rock salt mines will not be discussed in detail in this chapter. More on this subject can be provided by specialists (miners, geologists, geophysicists) active in potash and rock salt mining (e.g., EHRHARDT 1980). The same is true for all questions regarding mine safety in connection with gases and condensates.

An overview of the danger of gas and safety in potash and rock salt mining is given, for instance, by GIMM (1954), GIMM & ECKART (1968), GIESEL et al. (1989), and HEMPEL (1989).

Salt-bound gases can be detected above all with geological, mineralogical, and chemical criteria and geophysical methods of measurement.

Among the geological criteria are data on the attitude of potentially gas- and condensate-bearing beds as well as fractures and fault zones and a description of the geology of the rock surrounding and overlying the evaporites.

The mineralogical and chemical criteria primarily involve the relationships between the composition of the rocks and the occurrence of the gases, the type of fixation of the gases in the rocks, and questions regarding the formation and migration of gases in evaporites. The occurrence of salt solutions in the evaporites is to be considered for both the geological and mineralogical-chemical criteria for salt-bound gases.

According to W. RICHTER (1953) the determination of the dielectric constant should help in detecting incrystalline $CO_2$ assuming that solutions also occur together with the gases. It has been shown, however, that the method for predicting gas explosions is unsuitable (see ECKART et al. 1966: 148f).

In terms of geophysics, an attempt was made to localize the bed boundaries of gas-bearing rocks with reflection and refraction seismics (see GIMM & ECKART 1968: 55).

Due to its short, 20-m range the sound method with frequency-modulated sound is also unsuitable (in ECKART et al. 1966: 148).

An acoustic method for measuring gas based on the crackling sounds when gas is released from salts, particularly with heating and humidifying, is described and discussed, for instance, by WINTER (1964), ECKART et al. (1966: 155ff), and GIMM & ECKART (1968: 580f). Signs of salts posing a threat of explosion are also the degassing phenomena observed when drilling and the crackling sound of gas-bearing drilling powder.

The Freiberger gas-pressure prognosis involves measuring the increase in pressure at the borehole caused by degassing of the evaporites (GIMM & ECKART 1968: 582f). GIESEL (1968) suggested measuring the degassing rate at boreholes as volume over time. Even temperature changes in boreholes can be related to the gas content of the evaporite (GIESEL 1968: 106).

SCHRADER et al. (1960) developed a quick method for determining the gas content of evaporites by grinding samples in a vibration mill.

The Freiberger core prognosis is based on the observation that gas-rich drill cores fall apart in the form of hourglass-shaped segments (ECKART et al. 1966: 149ff; GIMM & ECKART 1968: 578f).

Mineral-bound gases are often detectable due to their smell.

It is not always possible to reliably predict gas releases with data of the gas contents in evaporites. One cause for this is surely that it is not possible to record the total amount of gases in cracks and fissures and in minerals simultaneously with one method. That is, information on the total volume of gas in an evaporite is difficult to obtain.

Furthermore, boreholes in gas-bearing evaporites do not always release gas. GIESEL (1968: 106) determined that deformation of rock to the extent that gas surrounding the borehole is released through cracks only occurs after a critical stress limit has been exceeded. In spite of this and other difficulties it is possible to minimize the threat posed by gas in potash and rock salt mines with analyses and safety measures.

Mining operations require the fastest gas measurement and decision-making possible. Hence, the geological and mineralogical study of gas-bearing evaporites which requires much time and technical expenditure is normally not suitable for immediate supervision of mining operations. Nevertheless, it is absolutely necessary that the geological and mineralogical-geochemical relationships of the occurrence and the distribution of salt-bound gases are known. Other data (e.g., geophysical) and observations of the occurrence of gases in evaporites can only be interpreted realistically when this basic knowledge is considered.

## 14.14 Research concept

Two research projects involving the geochemistry of gas-bearing evaporites of Zechstein sequences 1-4 are the basis for the concept described in the following (HERRMANN 1984b, 1986).

Research on the long-term safety of underground repositories can be described as follows:

### Concept

Marine evaporites frequently contain elements and compounds in the form of gas mixtures, condensates, and petroleum. This is true worldwide for the majority of evaporite deposits, independent of their stratigraphic position and composition (sulfate or chloride type). Only the composition and volume of the gases and condensates trapped in the evaporites show regional variations. The consequences with respect to mining of the gas-bearing and gas-barren evaporites, the construction of underground repositories, and the discussion on the genesis of the evaporites vary accordingly.

The evaporites of Zechstein sequences 1-4 in Germany (Upper Permian, sulfate with transitions to the chloride type) also contain gaseous components, in addition to solid and liquid ones. A portion of these were fixed along grain boundaries and in the mineral grains themselves during solution metamorphism, above all in the chloride rocks (mineral-bound gases). Other portions of gas (free gases) were obviously trapped in the fractures and fissures without accompanying aqueous solutions (fracture-bound gases). Free gases are also present in caverns.

In the Werra-Fulda mining district (Zechstein 1, central Germany) evaporites containing great amounts of gas occur at the level of the Thüringen and Hessen potash salt seams. These gas mixtures consist primarily of $CO_2$ and $N_2$ as well as minute amounts $CH_4$, $O_2$, $H_2$, Ar, He, CO, and $H_2S$. There is a danger of suffocation in a mine when concentrations of such gas mixtures are high.

The gas mixtures occurring in Zechstein 2-4 contain much greater portions of inflammable hydrocarbons ($CH_4$ and higher hydrocarbons) and $H_2$, in addition to $N_2$, $CO_2$, $O_2$, noble gases, and $H_2S$. Such gas mixtures with the appropriate composition and concentration are easily ignited in the mine by an open flame or spark.

Gases and condensates in the Zechstein evaporites have been known to occur since potash and rock salt mining began 125 years ago. In spite of the repeated discoveries of gas and petroleum during the exploration and mining of the evaporite deposits in the Werra-Fulda mining district (central Germany) and in Niedersachsen (northern Germany), hardly systematic studies on the mineralogy and geochemistry have been conducted to answer the following questions:

1. How much gas and condensate are stored in the various chloride rocks (e.g., rock salt, carnallitite, sylvinite, Hartsalz) and brittle rocks (anhydrite and argillaceous rocks)?

2. What are the individual components (quantitative) of the gas mixtures?
3. Are there any relationships between the fissures, fractures, and cracks observed in the horizontal and steeply inclined evaporite beds and the occurrence of gases, condensates, and salt solutions?
4. How can information on the source (and formation?) of the gas mixtures and condensates in the evaporites be obtained?
5. Can isotope studies of $CO_2$, hydrocarbons, and $H_2S$ provide supplementary information on the source of the gases?
6. Are there relationships between the composition of the gases, the chemistry of the solutions, and the mineralogy of the various evaporites?
7. Does the composition of mineral-bound gases (fixed in the presence of solutions) differ from that of fracture-bound gases (trapped in dry form)?
8. Do the gases and condensates fixed near the site of penetration into the evaporite (basalt dikes, marginal zones of salt domes) differ from the enrichments of salt-bound gases and condensates located deep within the evaporite?

   This question arises from observations of varying distances the mobile phases have travelled from their point of penetration into the evaporite to the site of reaction and storage. As the gases, condensates, and salt solutions migrated various distances through the evaporites, the composition of the gases may have changed due to the variable absorption of individual gas components in the aqueous solutions and saturated salt solutions.
9. Can criteria for predicting the occurrence of gas-bearing evaporites in parts of horizontal evaporite deposits and salt domes be derived from the geochemical-mineralogical findings of questions 1 - 8?

Results of equal significance to evaporite research, salt mining, and the evaluation of the long-term safety on underground repositories can only be obtained through comparative studies of the greatest possible number of individual salt-bound gases in Zechstein 1-4. Hence, the study of salt-bound gas must not be limited to any one of the localities mentioned above. For the evaluation of salt-bound gases in the salt domes of Niedersachsen, specific criteria from the occurrences of salt-bound gases in the salt deposits of central Germany are just as interesting as vice versa.

The research project can only be realized independent of the investigations accompanying underground exploration. In connection with questions regarding underground disposal the results are not only of interest in the exploratory stage of a body of salt, but also to evaluating the long-term safety of a repository for hazardous anthropogenic wastes.

## Previous studies

Studies on salt-bound gases in recent decades concentrated again and again on the question of how deposits with high gas contents can be recognized in time without triggering an explosion when driving tunnels and mining salt. The large gas occurrences with high $CO_2$ contents in the evaporites of the Werra-Fulda mining district (up to 25 m³ gas/t salt; WOLF 1965, quoted in GIESEL et al. 1989) were understandably the main object of study in this case. The occurrence of inflammable gases in the potash and rock salt mines were also studied. Both geophysical and geological-mineralogical studies were employed here. This research was conducted above all by the research group »Mineral-bound Gases«, which was founded in 1958 in the former GDR and worked on the scientific explanation of gas explosions in the GDR until 1969. In addition to geophysical investigations, this society also conducted mineralogical and chemical studies directed at solving problems in mining. One notable aspect of these studies was the concept of studies comparing the occurrence of salt-bound gases in the evaporites of the southern Harz mining district (primarily hydrocarbon-bearing mixtures of gas) and in the eastern part of the Werra-Fulda mining district (primarily $CO_2$-bearing gases). Practical suggestions for recognizing deposits with a high gas-explosion risk were the result of these studies (for summary see, e.g., GIESEL et al. 1989). The investigations on gas explosions in the potash and rock salt mines in central Germany were continued with great success in the 1970s and 1980s (e.g., GIESEL et al. 1989).

In the Federal Republic of Germany - on the west side of the former national border between the FRG and the former GDR - not as much data are available in the scientific literature as in the former GDR where data on the composition of salt-bound gases have been collected for 30 years. This fact is significant because knowledge on the salt deposits in the former GDR cannot be transferred directly to the evaporite occurrences in the FRG. Further and more extensive analytical and theoretical work is necessary, particularly on the Werra-Fulda mining district (western part) and in Niedersachsen (northern Germany).

The research project (»Salt-bound Gases in Zechstein Evaporites«) should contribute valuable information on this subject. In addition to providing a wealth of fundamental scientific research, this project will involve problems related to the exploration of evaporites, salt mining, and the underground disposal of anthropogenic wastes in evaporites. If there are ways of predicting deposits with a high danger of gas explosions and the source of the gases, they can only be discovered using the results of fundamental geochemical research on the relationships between solid, liquid, and gaseous components on the evaporites of Zechstein sequences 1 - 4. Such research has not yet been conducted.

## Aims

1. A data basis is to be established for the volumes of gas contained in the evaporites. The studies must give equal consideration to the rocks of Zechstein 1 (Werra-Fulda mining district) and Zechstein 2 - 3 (northern Germany; e.g., KNABE 1989).
2. A data basis is to be established for the quantitative composition of the gases contained in the evaporites. It is to be checked, for example, whether there are any

relationships between the composition of the gases (and condensates) and the reservoir rock (e.g., KNABE 1989).

3. If possible, the distribution of gas in the individual reaction zones in the spatially restricted metamorphic areas should be investigated with respect to the mineralogical and chemical composition of the various zones of alteration, the quantity of reaction solutions, and the temperature distribution around sources of heat. Suitable objects for study are the basalt intrusions and surrounding evaporite rocks (e.g., rock salt) in the Werra-Fulda mining district. Here, the effects of disposing of heat-generating wastes in gas-bearing rock salt can be studied using a natural occurrence (natural analogue).

4. An attempt should be made to determine the composition and volume of gases stored in cracks and fissures, in addition to those stored in minerals. Since fracture-bound gases apparently migrated to their present site of storage in dry form (i.e., not accompanied by aqueous solutions), there is a chance that the original composition of the gases which penetrated into the evaporites has remained nearly unchanged.

5. Additional information on the source of the gases and condensates can probably be obtained with isotope studies ($\delta^{13}C$, $\delta^{34}S$) of the C and $H_2S$ components of the gases.

6. When interpreting gas compositions, the transport distance through the evaporites is to be given the same consideration as the differential absorption of the individual gas components in aqueous solutions and saturated salt solutions of differing composition (e.g., NaCl or $MgCl_2$ saturated, Bunsen absorption coefficient).

7. Considering item 6, the composition of mineral-bound gases in rocks intensively altered by solution metamorphism can differ from that of gases in rocks which experienced a less intense metamorphism. Differences between fracture-bound gases and those near former heat sources (e.g., basaltic melts) are also conceivable. These are the starting points for the practical application of geochemical gas analyses in salt mining for predicting gas-bearing parts of evaporite deposits (see KNABE 1989).

**Part III**

**The Composition of Salt Domes as a Criterion for Evaluating the Long-Term Safety of Underground Repositories for Anthropogenic Wastes**

# 15 Geological principle

Before underground repositories for anthropogenic wastes in bodies of rock (i.e., geological systems) are constructed, one question must be asked regarding every prospective site (see also Part I): What events can adversely affect the long-term safe disposal of the wastes? Such events involve geological time, extending thousands to over one million years into the future. During this time substances in harmful concentrations must be prevented from leaking out of the repositories and reaching the biosphere.

The evaluation of the long-term safety of repository sites requires in-depth scientific study of geological systems which are already millions of years old. In the course of geological time the original spatial configuration and composition of these systems have been altered extensively many times. By modelling the changes that have occurred, prognoses of the probable evolution of the repository rock in the geological future can be made. This procedure is comparable with putting together a complicated mosaic.

The long-term safety of a geological repository system is mainly influenced by natural processes, independent of man. Among these are geological events such as regional uplift and subsidence as well as volcanism, frequently accompanied by earthquakes. However, the influence of such events on the integrity of underground repositories can surely be disregarded in the near geological future. In contrast, much greater importance has to be placed on the chemical and physical processes which have been effective since the earth's crust formed and which can change the composition of the rocks locally or over vast areas. All interactions between mobile components (fluids) and the solid minerals of the rocks are meant here.

The fluid components consist of compounds and elements such as water, hydrogen chloride, carbon dioxide, sulfur dioxide, hydrogen, nitrogen, oxygen, sulfur, and others which are in a liquid and/or gaseous state. Fluid components have differing origins, e.g., magmatism, formation waters (interstitial solutions), groundwaters, and surface waters.

Interactions between fluid components (mostly in the form of aqueous solutions) and solid phases are known to occur in all igneous, metamorphic, and sedimentary rocks. Their existence is proven by certain mineral reactions and the partial dissolution of minerals, i.e., by changes in the original mineralogical and chemical composition of the rocks. The fact that substances which are dissolved (mobilized) underground are able to reach the biosphere is important to all investigations on the long-term safety of repositories. There is frequently a connection between the uplift or subsidence of rock and the intensity of interactions between fluid components and the rocks.

The effect of fluids on rock composition is currently the object of intensive study. Fluids are also of great interest to problems pertaining to repositories. *In the near*

*geological future, fluid components are the only factor able to reduce the effective-
ness of the natural and man-made barriers between an underground repository and
the earth's surface to the extent that the biosphere is threatened.* In this context the
various possibilities for chemical reactions between the radioactive wastes themselves
or with the repository rock as well as the aforementioned earthquakes and magmatism
play equally subordinate roles. Hence, in the following the relationships between the
effect of fluid components in geological systems (particularly evaporites), the material
transport caused by the fluids, and the long-term safety of repositories for radioactive
wastes (e.g., salt domes) will be in the limelight.

We must proceed from the following two facts:

1. *Geological systems are not in a static, but a dynamic state. This means that all
   bodies of rock (and the waste repositories therein) participate in chemical cycles
   and are in an everchanging state.*

2. *Prognoses of the long-term safety of underground repositories require long-term
   observations and experiments. Nature is the only laboratory for such studies. In
   nature, the results of both slow and rapid mineral reactions and material transport are documented (i.e., preserved). In man-made laboratories, on the contrary,
   only short-term experiments can be conducted and the results extrapolated for
   longer periods of time. Therefore, conclusions on long-term safety must also be
   backed up by observations in nature and cannot be based on laboratory experiments alone.*

# 16 The example of evaporites

In the discussion on the construction of underground repositories and their long-term safety the argument is occasionally put forward that scientific research should not study these problems for the sake of research alone: Current knowledge cannot be applied to the evaluation of long-term safety when the conditions under which the underground repositories were constructed no longer exist in the future, e.g., due to unpredictable climatic changes and crustal deformation.

The concept for evaluating the long-term safety of underground repositories presented here conflicts with this and similar theories in one basic aspect. We are not speaking here of research for the sake of research itself: We must make every responsible effort to apply current scientific knowledge to the central theme of environmental protection with the intention of adapting safety measures to the processes observed in nature.

Research on the topic of composition and long-term safety is not aimed at predicting what geological events will occur, when, and what effects they will have on the integrity of an underground repository in a natural body of rock. The scientific basis of this research lies in obtaining data on all the effects various geological events had on evaporites through the study of 250-million-years-old salt rocks (e.g., the Gorleben salt dome). In this way it is possible to study various site-specific geological and geochemical processes together with the physical and chemical effects they might have on the composition of the evaporite deposits of Hessen and Niedersachsen (Germany).

VENZLAFF (1978) pointed out the various geological events that affected German evaporites during their 250-million-years existence. The essential characteristics of the dynamic development of German Zechstein evaporites are given in a simple summary in Tab. 10. The dynamic history of the salt domes in Niedersachsen can be characterized by two events: first, the formation of the evaporites 250 million years ago in a region to the south at 30°N lat., where seawater evaporated continuously under the given climatic conditions (evaporation > precipitation; see, for example, GORDON 1975), and later, reached their present position much farther north as a result of plate tectonics. For example, the evaporites of the Gorleben salt dome are located today at 53°N latitude and 11°20'E longitude (Tab. 10, Fig. 19). The second event involved the deformation of the salt deposits and the development of salt domes.

The following processes and their consequences are always dependent upon the depth in the crust, at which the underground repository is to be constructed. Depths normally range from one hundred to over one thousand meters.
- What consequences have to be reckoned with in the case of an ice age in the future when glaciers physically stress the subsurface and melt waters cause subrosion?
- What effects would a marine transgression have?
- What effects would a change to an arid climate have?

**Tab. 10** Dynamic development of the Zechstein evaporites and the neighboring rocks during the last 250 million years. The alteration of different salt domes did not necessarily occur simultaneously. The data of the columns Overlying rock and Evaporites refer to the Gorleben salt dome (VENZLAFF 1978, JARITZ 1980, own data).

| Era | Formation | million years | Developments and changes | | | |
|---|---|---|---|---|---|---|
| | | | Plate tectonics | Earth's surface | Overlying rock | Evaporites |
| Cenozoic | Quaternary | 1.8 | Present situation | Continental conditions; ice ages | The overlying rock is stressed by ice ages | Subrosion in the upper part of the salt dome. Alterations and material transport |
| | Tertiary | 65 | Dereasing activity of crustal movements | Terrestial-marine changes. Tropical climate | Younger sediments overlie the salt dome | Further development of the salt dome. Mineral reactions, material transport? |
| Mesozoic | Cretaceous | 141 | Intense crustal movements | Terrestrial-marine changes | Evaporites penetrate into overlying sediment strata | Formation of the salt dome. Subrosion? Fracturing, material transport |
| | .Jurassic | 195 | Beginning of crustal movements Evaporites > 30°N | Terrestrial-marine changes | Continued updoming of overlying rock | Further salt migration into the pillows. Initial salt dome formation |
| | Triassic | 230 | Evaporites < 30°N | Terrestrial-marine changes | Updoming of overlying rock | Initial deformation of salt layers. Formation of salt pillows |
| Paleozoic | Permian | 280 | Pangea. Formation of evaporites south of < 30°N | Marine conditions during upper Permian | No Overlying rock | Crystallisation of evaporites during upper Permian (about 250 million years b.p.) |

- What effects would a change to a tropical climate have?
- What would be the effects of renewed evaporite deformation with the same intensity as that during salt dome formation (Jurassic, Cretaceous)?
- What consequences would have to be reckoned with in the case of volcanism with basaltic intrusion into the evaporite of the geological repository system as happened 17-25 million years ago in Hessen (Germany)?
- What would be the effects of earthquakes accompanying volcanism in the evaporite bodies?
- What effects would renewed faulting of the continental crust have?

The aforementioned events cover all essential processes which could actually or theoretically influence the original mineralogical and chemical composition of an evaporite body or the integrity of a repository for radioactive or nonradioactive wastes in the future.

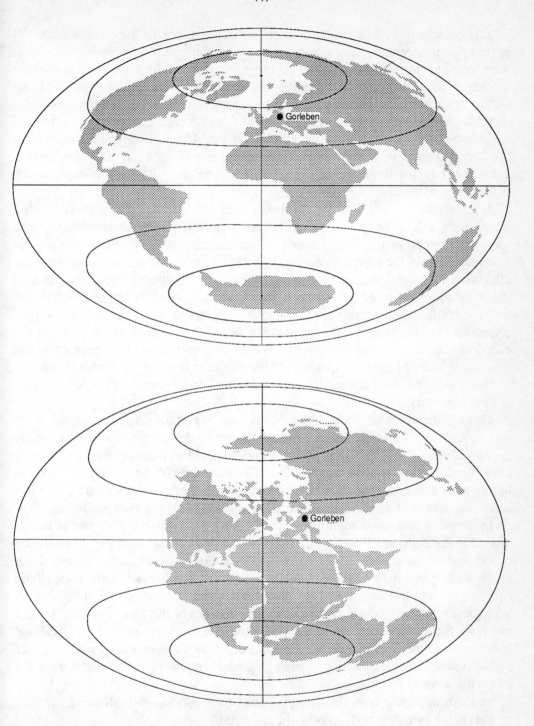

**Fig. 19** Change in the geographic position of the Zechstein evaporites of central and northern Germany from < 30°N latitude to 53°N latitude since their formation 250 million years ago (projection of continents after Smith & Briden 1977)

It is obvious that current conditions for protecting man (not the entire biosphere!) against anthropogenic wastes would no longer be fulfilled if such geological events were to occur. In this case it would no longer be possible for humans to live in the vicinity of a repository site in the current numbers and with the present quality of life. On the other hand, the effects of natural processes can also serve as examples of the possible changes which man can trigger in a body of rock: e.g., salt domes subjected to stress by high-heat-generating radwaste which could have effects similar to those of spatially and temporally limited processes during dome formation 100 - 150 million years ago (e.g., the formation of paths for the migration of fluid components during deformation of the evaporite bodies).

It is easy to see that it is not absolutely necessary to know the time at which a future event will occur for evaluating the long-term safety of an underground repository. Based on present knowledge such predictions are only of speculative nature and will remain so in the future. However, it is important to know something about the effects the aforementioned events and related physical and chemical processes have on the composition of the repository rock, which material transport can occur under the influence of fluid components, and how to evaluate quantitatively their possible consequences. In rare cases it may be possible assign the phenomena observed in the evaporites and in the surrounding rocks to a specific geological event or process. But the change in the original composition of the evaporite bodies by a sum of geological events is also of great interest in the evaluation of the long-term safety of an underground repository.

Considering all observations it can be determined that the original composition of natural rocks has been altered to varying degrees in the course of several or many geological events. However, it is also conceivable based on these observations that the mineralogical and chemical composition of certain parts of the repository rock has hardly been changed, if at all, since its formation and in spite of various geological processes over millions of years. Research on this subject is just now beginning.

In Germany hazardous anthropogenic wastes have mainly been disposed of underground in salt deposits. This disposal practice will probably be continued in the future (using other rocks as well). Since aqueous salt solutions are practically the only threat to the long-term safety of repositories in evaporites, the interactions between fluid components and evaporites will be discussed in this context in the following. A geological repository system in evaporite rock consists of the rock formation (e.g., a salt dome), the repository, and the surrounding and overlying rock (Fig. 20). Fluid components in the form of aqueous solutions occur both within and outside evaporite bodies (see also Part II). That is, aqueous solutions can effect an evaporite body in the following ways:

1. solutions migrating from the overlying and/or surrounding rock affecting the margins of the evaporite body, especially subrosion,
2. solutions migrating from the overlying and/or surrounding rock and penetrating into the evaporite body, and

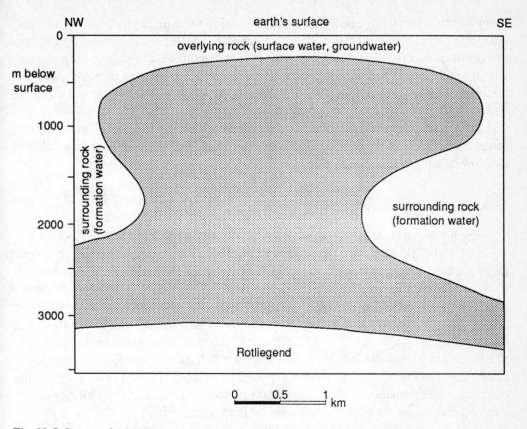

NW                                    earth's surface                                    SE

overlying rock (surface water, groundwater)

surrounding rock (formation water)

surrounding rock (formation water)

Rotliegend

0     0.5     1 km

**Fig. 20** Influence of solutions on a salt dome (from HERRMANN & KNIPPING 1988a).

3. solutions trapped within the evaporite body (e.g., in anhydrite or rock salt beds) which can be remobilized by rock movement; the volume of such solutions can range from several liters up to several thousand cubic meters (e.g., HERRMANN 1983a: 175).

## *How do such fluid components spread through evaporites?*

Porosity and permeability are criteria for fluid movement within rock.

Distinctions are made between intergranular (voids between mineral grains) and incrystalline or ingranular (voids within the grain) porosity as well as between porosity in the form of fractures and fissures (caused by mechanical and chemical processes) and caverns (caused by subsequent chemical processes). There are similar differences between various permeabilities (e.g., SCHOPPER 1982: 185, 282). Porosity and permeability are dependent on various factors: e.g., the composition of the rocks, grain shapes, grain size, and pressure exerted on the rock.

In chloride rocks (especially rock salt) the paths for liquids produced by intergranular and incrystalline porosities and permeabilities are generally limited. In the case of porosities and permeabilities contingent upon fractures and fissures, there are numerous indications that cracks, joints, fractures, and fissures are possible paths for liquids and gases in evaporites. Such features occur mainly in anhydrite and salt clay, and more rarely in rock salt and potash salt beds.

In contrast to the evaporite rocks, the sedimentary rocks (frequently limestone and sandstone) enclosing the evaporites do have porosities and permeabilities favorable to the flow of aqueous solutions.

## How do fractures in evaporites form?

The flat layered evaporite deposits of the Werra-Fulda mining district (Hessen) at a depth of 500 - 700 m and the salt domes of Niedersachsen extending down to several thousand meters are equally suitable objects of study for answering this question.

The Zechstein evaporites which formed about 250 million years ago (MENNING 1986) were intruded frequently by basaltic melts 17-25 million years ago. These basalt dikes ascended from the earth's mantle within several hours and days and solidified in the evaporites (Fig. 21). These intrusions produced stresses in the evaporite rocks, leading to the formation of fractures and fissures (hydrofracturing). At the same time, basaltic melts and fluid components (aqueous solutions and gases) penetrated into these fractures. The composition of these basalts and their effects on the intruded evaporites are a unique object of study (KNIPPING 1989)

Fractures, due to quite different geological events however, also occur in the salt domes of Niedersachsen. These fractures are attributed to deformation triggered by halokinesis and tectonics in the subsurface and led to the formation of salt domes.

## How are fractures as paths for fluids to be evaluated?

It has been observed both in flat layered salt beds and salt domes that cracks, gaps, and fissures usually reheal due to the plastic behavior of evaporites such as rock salt, or are filled with secondary minerals such as halite, sylvite, kainite, carnallite, and others. Secondary mineral fillings have been observed frequently in all known evaporite bodies. The sealing of former paths for mobile components due to plastic deformation or secondary mineralization is so complete in most cases that they can no longer serve as preferred paths for solution and gas migration. However, this can change with renewed deformation of the evaporite beds. Such deformation does not necessarily have to be triggered by geological events. It can also result from activities related to underground repositories. Of the different kinds of radioactive and nonradioactive wastes, however, only the high-heat-generating (high-level radioactive) substances can produce clearly measurable deformation of an evaporite body. Such deformation can be neglected for wastes which generate no or only small amounts of heat.

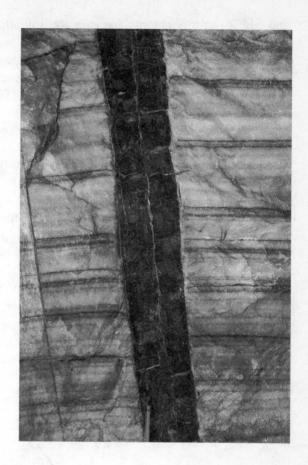

**Fig. 21** Basalt dike in the middle Werra
rock salt (Na1β), Hattorf mine
of Kali & Salz AG (from
KNIPPING 1989)

Regarding the overall age of the Zechstein evaporites (about 250 million years) and their volume it is obvious that fractures in the evaporite beds only acted as potential paths for fluid components during deformational events, i.e., temporally and spatially restricted. During deformation, however, there were, in part, extensive mineral reactions and material transport in greater portions of the individual evaporite bodies.

## How do aqueous salt solutions and gases react with the enclosing evaporite rocks?

Mobile components, particularly unsaturated aqueous solutions, react with the individual evaporite rocks in different ways. The reason for this is the differing solubility of the alkaline and alkaline-earth sulfate and chloride compounds in water. Of the minerals occurring in evaporite sequences, $CaCO_3$ (calcite), $CaMg(CO_3)_2$ (dolomite), and $CaSO_4$ or $CaSO_4 \cdot 2H_2O$ (anhydrite, gypsum) have the lowest solubilities. NaCl and MgCl$_2$ only negligibly increase the solubility of these compounds in aqueous solutions.

This is not true for the chlorides of the evaporite sequences, i.e., halite (rock salt, NaCl) and minerals of the K-Mg rocks (potash salts) such as sylvite (KCl), carnallite (KMgCl$_3$ · 6H$_2$O), and kainite (KMgClSO$_4$ · 2.75H$_2$O). These chlorides are the most soluble compounds in nature. Nevertheless, the rock salt and potash rocks behave differently under the influence of mobile components in the form of unsaturated aqueous solutions, as nature has demonstrated time and again. Findings will be discussed in the following with one example from flat layered salt beds and one from a salt dome.

Fig. 22 shows the effect of fluid phases (aqueous solutions and gases) from a basalt dike on the flat layered rock salt beds separated in the middle by a potash salt seam (cf. Fig. 17). Such stratigraphic relationships can be studied in the Werra-Fulda mining district.

The basaltic melts had a temperature of 1150 °C and the evaporite rocks a temperature of about 50 °C before intrusion of the magma. The basaltic melts were accompanied by mobile constituents, which had a temperature of only one hundred to several hundred degrees Celsius. The study of the evaporite rocks clearly shows that the alteration of the rock salt by the fluid components only extends a few centimeters into the evaporite and away from the contact to the basalt. In contrast, the mobile compo-

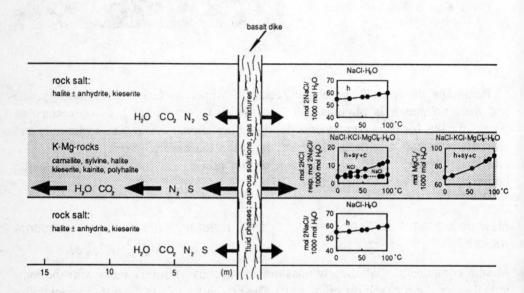

**Fig. 22** Various effects of fluid phases of the basaltic volcanism on evaporites of differing chemical composition. Werra-Fulda mining district, Germany, Zechstein 1 (Werra sequence), as an example of flat layered evaporite beds.

nents penetrated much deeper into the K-Mg mineral associations of the potash salt seams, producing a zone of alteration which is up to and over 10 m wide.

These observations can be explained by the differing solubility of halite, (NaCl), carnallite (KMgCl$_3$ · 6H$_2$O), and sylvite (KCl) and by the volume reduction of the potash salt seam caused by the dissolution of the rock. Between 50 °C and 100 °C the solubility of NaCl in the NaCl-H$_2$O system increases by about 8 mol NaCl/1000 mol H$_2$O. This means that aqueous solutions are already saturated with NaCl in the immediate vicinity of the basalt-evaporite contact and then are obviously not able to penetrate further into distant rock salt beds due to the low porosity and permeability (presuming the absence of fractures!). However, this is not true for KCl and MgCl$_2$ compounds in the NaCl-KCl-MgCl$_2$-H$_2$O system. In this system, although the solubility of NaCl remains practically unchanged between 50 °C and 100 °C, the solubility of KCl increases by about 9 mol KCl/1000 mol H$_2$O and that of MgCl$_2$ even increases by about 15 mol MgCl$_2$/1000 mol H$_2$O. Thus, if the mineral carnallite, for example, was present in the potash salt seam before the mobile components intruded, aqueous solutions and gases would have been able to spread much further there due to the increasing solubility of KCl and MgCl$_2$ with increasing temperature and the volume reduction of the original rock. Similar conditions are found in the salt domes of Niedersachsen although the initiating events were quite different.

Fig. 23 shows an outline of the Gorleben salt dome where the prototype of a repository for radioactive wastes is to be constructed if the evaporites proves suitable. Previous surface reconnaissance has shown that steeply inclined potash salt seams (Staßfurt seam, K2) and anhydrite beds (Haupt- and Leine anhydrite, A3) extend up to the cap rock. The chemical and mineralogical composition of the potash salt seams and enclosing rock salt was able to be studied using drill cores. It was discovered that the composition of the potash salt seam from top to bottom has obviously been more intensively altered than the neighboring rock salt. Unsaturated solutions from the overlying rock penetrated into the potash seam of the Gorleben salt dome. The most intensive alterations and material transport occurred in the upper zone (approximately 90 - 130 m below the surface of the salt wash surface in the range of a Quaternary channel, drillings GoHy 1301 - 1305). All potassium minerals were dissolved, leaving mostly rock salt behind. In the underlying zone, the original mineral association of kieserite and carnallite (MgSO$_4$ · H$_2$O + KMgCl$_3$ · 6H$_2$O) was converted into kainite (KMgClSO$_4$ · 2.75 H$_2$O; 140 - 170 m below the salt wash surface, between the drillings GoHy 1304 and 1305). The first nearly unaltered carnallitic rock is found below (e.g., TÄTIGKEITSBERICHT DER BGR 1985-1986: 56f; BORNEMANN et al. 1988).

In spite of the very different sources of the fluid components (magmatic origin in the Werra-Fulda mining district and penetration from the underlying rock; surface and formation waters for the salt domes in Niedersachsen and penetration from the hanging-wall rocks), the mineral reactions, the material transport, and the differing depth of penetration in the potash salt and rock salt beds are comparable.

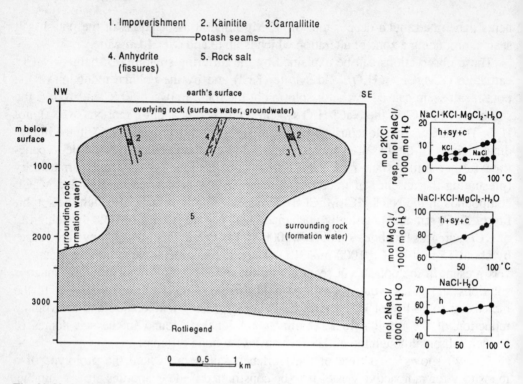

1. Impoverishment   2. Kainitite   3. Carnallitite
                    └──────Potash seams──────┘

4. Anhydrite        5. Rock salt
   (fissures)

**Fig. 23** Various effects of fluid phases from the overlying rock of a salt dome (example Gorleben, Niedersachsen, Germany) on rock salt and potash salts (potash seam Staßfurt, K2). Inside the salt dome mineral reactions and material transports reach deeper into the potash seams compared with the rock salt. Example of steep layered evaporite beds.

*The necessity for comparative studies of the various evaporite occurrences regarding their use as underground repositories for anthropogenic wastes cannot be illustrated in a more impressive way.*

## At what rates do solutions migrate through evaporites?

This is a difficult question to answer. Based on observations and comparisons it appears conceivable that solutions can migrate over distances of up to 10 m and more within an evaporite body not in millions of years, but in considerably shorter periods of time. This applies particularly to the fluids invading under pressure, such as those which accompanied the basalts in the Werra-Fulda mining district (e.g., KNIPPING & HERRMANN 1985, HERRMANN & KNIPPING 1988a).

## Preliminary result

Potential paths for mobile components (aqueous solutions) in evaporite beds (both flat layered and vertical) are due to the solubility of their minerals the potash salt seams and due to fissures anhydrite and salt clay and sometimes rock salt. A representative example of the latter case is the occurrence of fissures up to 10 cm in width in the hanging-wall salt of the Staßfurt sequence in the Gorleben salt dome. Such fissures were found in core material of drilling Go 1005 below 421.7 m (126 m below the salt wash surface). Fissures are closed in part with cm-size halite crystals. Carnallite is found occasionally as well (e.g., FISCHBECK & BORNEMANN 1988).

In the case of flat layered salt beds (e.g., the Werra region) the effect of aqueous solutions (formation and surface waters) on the potash salt seams is presently to be expected above all at the margins of evaporite deposits (salt table). No formation water penetrates into the salt deposits from above since they are well protected by several meters of clay and other strata. Additionally, there are presently no known signs of renewed magmatism in the Werra-Fulda mining district. This means that underground repositories have to be constructed in portions of the sequence which are as far away as possible from the salt tables (e.g., Herfa-Neurode).

In contrast, steeply inclined evaporites (salt domes, e.g., in northern Germany) have to be evaluated in a different light. It has to be assumed here that aqueous solutions will penetrate into potash salt seams and beds of anhydrite from above. *Hence, underground repositories for hazardous wastes must be situated as far away as possible from such strata and at greater depths under the earth's surface and the salt wash surface within the rock salt of a dome.*

## The occurrence of solutions in potential repository salt domes

It is presumed that sound prognoses of the long-term safety of underground repositories are only possible when all conceivable effects of solutions on geological systems are taken into consideration. In the case of the Gorleben salt dome, the rocks overlying the salt dome has been the primary object of study regarding the occurrence and distribution of solutions. Such studies showed that portions of the upper part of the salt dome have been removed by subrosion during the Quaternary (approximately 0.1 - 0.8 million years; e.g., ZUSAMMENFASSENDER ZWISCHENBERICHT DER PTB 1983: 53). Today, the contact between the salt and cap rock is situated at an average depth of about 250 m. The zones of alteration in the Staßfurt potash salt seam illustrated in Fig. 23 were also formed by subrosion.

Using model calculations, EHRLICH et al. (1986) estimated how much time subrosion from outside the evaporite body would need to completely dissolve the approximately 500-m-thick evaporite beds (mainly rock salt) between cap rock and repository (850 m deep). Assuming ascent rates of 0.01 - 1 mm per year, 50 - 0.5 million years would be necessary. These data are based on a theoretical worse-case model. The selective subrosion of a salt dome out of the surrounding rock down to depths of

800 m and more has never been observed in nature. Geologically, however, it is conceivable that local fluid movement is possible within the salt dome upon renewed deformation of the evaporite.

Due to the lack of outcrops little can presently be said as to whether and to what extent mobile components have altered the original composition of the Gorleben salt dome in the geological past. To date, only four deep drillings on the flanks and two shaft drillings in the center of this salt dome have been sunk. However, knowledge of the solutions and above all of potential paths for mobile components in the evaporites can be improved substantially with subsurface exploration. At the present we can rely on the following founded knowledge:

In *all* previously mined salt domes of Niedersachsen (potash and rock salt mines) solutions have been released from evaporite beds over the last 100 years. The Gorleben salt dome is no exception.

Fluids have been released in all four deep drillings sunk into the flanks of the Gorleben salt dome (e.g., ZUSAMMENFASSENDER ZWISCHENBERICHT DER PTB 1983; Fig. 24). There were seven releases of fluids from the salt dome, one of which was from the surrounding rock (drilling 1005). The following three observations are noteworthy:

1. The sites of release and the possible reservoir rock for the solutions were the Leine anhydrite (Hauptanhydrite, A3) in six cases. Only in one case the solutions was released from the uppermost Staßfurt rock salt (Na2) directly below the Staßfurt potash seam (K2).

2. The solutions were released from points distributed over a depth range of about 2000 m.

3. All solutions were concentrated salt solutions in which the $MgCl_2$ portion dominated having 290 - 440 g/l. The analysis of one such solution is given in Tab. 11 (see ZUSAMMENFASSENDER ZWISCHENBERICHT DER PTB 1983: 39; HERRMANN 1984a: 443; HERRMANN & KNIPPING 1989: 13).

It is not sure whether the volumes of the solutions released in the boreholes are representative of the total solution stored. Based on experience with salt deposits of the German Zechstein volumes on the order of 1 - 1000 $m^3$ are possible.

The findings from Gorleben and other salt domes in northern Germany agree with known observations of the paths for liquids and gases in evaporites and the behavior of mobile components in evaporites. Six of the seven solution releases in the Gorleben salt dome were from the partially fractured Leine anhydrite (Hauptanhydrite, A3). In addition, the dominant interaction between the aqueous solutions and the rocks of the potash salt seams is expressed in the chemical composition of the salt solutions from the drillings (Tab. 11). These salt solutions are obviously not concentrated seawater from the former depositional basin and not solutions which formed during the dissolution of minerals *outside* the salt dome. Apparently, the high amount of $MgCl_2$ detected in the solutions can only be explained by the influence of aqueous solutions on the particularly reactive carnallitic rock of the Staßfurt potash seam in the Gorleben salt dome.

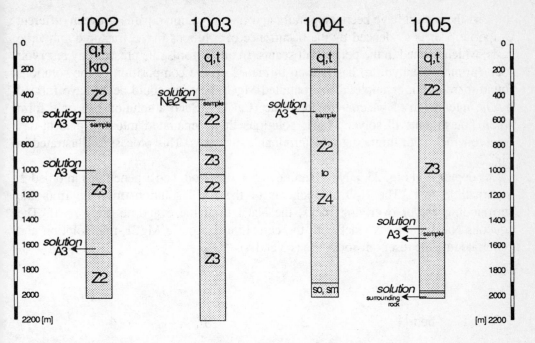

**Fig. 24** Profiles of deep drillings 1002-1005. Arrows indicate points of solution release. Z1, Zechstein 1, etc.; Na2, Staßfurt rock salt; A3, Leine anhydrite (Hauptanhydrite); sm, middle Buntsandstein; so, upper Bundsandstein; kro, upper Cretaceous; t, Tertiary; q, Quaternary (from HERRMANN 1984a; 442).

**Tab. 11** Composition of solution from Gorleben deep drilling 1003 from a depth of 431-449 m. Solution with density of 1.312 g/cm³ at 20°C (analysis by Preussag AG 1980).

| components | mass fraction in % | g/l |
|---|---|---|
| NaCl | 0.97 | 12.7 |
| KCl | 0.51 | 6.7 |
| MgCl$_2$ | 29.2 | 383.7 |
| MgSO$_4$ | 0.77 | 10.1 |
| CaSO$_4$ | 0.18 | 2.4 |
| H$_2$O | 68.4 | 896.4 |
| Br | 0.535 | 7.02 |
| Li | 0.0050 | 0.066 |

That there must have been temporally and spatially limited paths between different evaporite beds is evidenced by the occurrence of solutions in the fractured anhydrite beds, which formed in the potash salt seams. In other words, the present-day reservoir rock (primarily anhydrite) has had no influence on the composition of the solutions found therein. For example, a concentrated $MgCl_2$ solution could never have formed by the interaction of water with anhydrite ($CaCO_4$). Such a solution can only form when potash salts dissolve. $MgCl_2$ solutions then penetrated into the present-day reservoir rock after migrating undeterminable distances. This process is illustrated in Fig. 25.

As depicted in Fig. 25, a NaCl-saturated solution from bed 1 penetrates into bed 2, a carnallitic rock. The $H_2O$ components of the NaCl solution originate from the surrounding and/or overlying rocks, the NaCl from the evaporite body itself. The aqueous NaCl solutions react with the carnallite to form a $MgCl_2$-rich solution and penetrate into the reservoir rocks (beds 3 and 4).

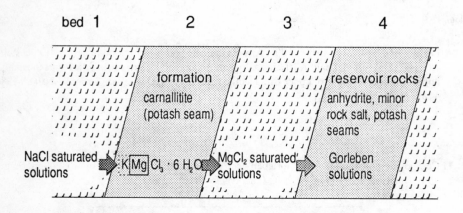

**Fig. 25** Schematic representation of the formation of dominantly $MgCl_2$-bearing solutions in a salt dome and the possible paths into the present reservoir rocks. The significance of beds 1 - 4 is explained in the text (from HERRMANN 1984a: 445).

Solutions are obviously able to migrate most easily when evaporite beds are deformed or when liquids penetrate into the rock under pressure (see fluid components of the basalt magmatism in the Werra-Fulda mining district). Only the first case must be taken into consideration when constructing repositories in evaporites.

The following remarks should show how important intensive evaporite research - involving as many different salt deposits as possible - is to the subject long-term safety of underground repositories in evaporites. Only in this way the detailed information on the effects of fluid components on evaporites can be obtained which is necessary for planning underground repositories. An attempt will also be made to show how vast the subject area of salt solutions in evaporite research is.

*Theoretical knowledge is needed for properly managing all information - extensive knowledge of such deposits and experience with controlling releases of solutions in potash and rock salt mining is just as important .*

## Summary of the knowledge on long-term safety

How can the previously discussed details be summarized and meaningfully expressed in statements on the long-term safety of underground repositories in salt domes? The method currently being tested in the Gorleben salt dome (applicable to other salt domes and nonradioactive hazardous wastes) is illustrated in Fig. 26.

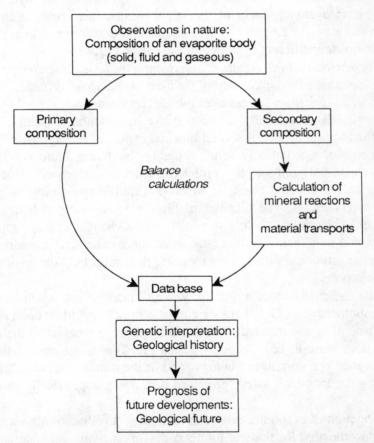

**Fig. 26** Principle for quantifying mineral reactions and material transports in evaporites to be applied to assessing the long-term safety of repositories for anthropogenic wastes (after KNIPPING 1988)

The principle is based on the quantification of all previous changes in the original composition of the evaporites related to mobile components. The primary aim here is the estimation of the quantities of solutions involved in the mineral reactions and material transport. This is done by means of complicated geochemical criteria. In this context it might be possible to delimit the zones within a salt dome which obviously did not come into contact again with fluid components after their precipitation from seawater about 250 million years ago (e.g. HERRMANN & v.BORSTEL 1991). To our knowledge, the composition of the various salt domes in northern Germany has been influenced by aqueous solutions to varying degrees in the geological past. This can be recognized above all in the mineralogical composition of the potash salt seams which reacted in a particularly sensitive way to the effect of unsaturated solutions, as has been demonstrated. Hence, the potash salt in the salt domes of northern Germany are - in addition to rock salt - significant as an index of the changes in the original composition of the evaporite deposits.

If, for example, the potash salts within the evaporite body (at depths of 800 - 1000 m) remained essentially unchanged in spite of the unfavorable hydrogeological conditions in the overlying rocks of the salt domes, it must be carefully checked whether the salt acting as a barrier is more effective in the long term than the caprock which has completely different hydrogeological characteristics.

*We are optimistic that the results of safety analyses based on other prerequisites can be supplemented with our concept of the dynamic nature of geological repository systems and by considering changes in evaporite composition.*

The objection is occasionally raised that the construction of shafts, drifts, and repository chambers at depths of several hundred meters adversely affects the closed system of a body of rock not only during operation, but in the future as well. For this reason it is supposedly impossible to infer the processes which could occur in the future based on the material transport which occurred in the geological past under closed-system conditions. The disposal of high-heat-generating radioactive wastes would also influence the geological system: Previous changes in the composition of evaporites would lose their informational value for geochemical assessment of the processes in the geological future. This argument disregards both the engineering and geoscientific aspects.

The rooms and chambers of a repository mine - planned and constructed specifically for this purpose (e.g., Gorleben) - are to be completely refilled upon completion of disposal activities, and the shafts - as the only possible accesses to the repository from the surface - are to be effectively sealed off. This is possible  based on the experience gained and current technology used in the mining industry. The concept for evaluating the long-term safety presented here applies specifically to repository mines.

The deformation of evaporites related to heat (high-level radioactive wastes) does not overshadow the need for study of former paths for solutions and material transport in rocks. On the contrary, the formation of potential paths related to evaporite deformation (e.g., during salt dome formation) emphasizes the necessity for the study of compositional changes caused by aqueous solutions in a salt dome in the geological past.

For the purpose of this work, the changes in original composition which have occurred in evaporites are to be recognized and quantified first. By doing so, information not only on the previous extent of the compositional changes in the salt dome, but also on possible paths for aqueous solutions in the past can be obtained. The value of information for evaluating the geological barrier effect of a repository mine is obvious. In a second step an attempt will be made to infer information on processes which might occur in the future based on findings on the previous development of a geological repository system.

*The concept for studying the changes in the mineralogical and chemical composition of a geological repository system presented here is an indispensible part of every geologically founded statement on the long-term safety of underground repositories for hazardous anthropogenic wastes.*

# 17 Composition of Zechstein evaporites in the Hannover region, northern Germany

## 17.1 Evaporite rocks

The evaporite bodies in the Hannover region are composed primarily of evaporites of the Zechstein 2-4 (the Staßfurt, Leine, and Aller sequences). There are also occasional occurrences of non-Zechstein (mostly Triassic) evaporites, particularly on the flanks of salt domes.

The stratigraphy and composition of the evaporites of the Gorleben salt dome have been studied by means of exploratory drillings, shaft drillings, and salt table drillings (BORNEMANN & FISCHBECK 1986; BORNEMANN et al. 1988; BORNEMANN & FISCHBECK 1989; BORNEMANN et al. 1989).

Besides the predominant Staßfurt, Leine, and Aller rock salt (Na2-4) and like nearly all evaporite bodies in the Hannover region, the Gorleben salt dome is also made up of salt clays (Gray salt clay or lower Leine clay, T3; upper Leine clay, T3r; red or lower Aller clay, T4), carbonate (Leine carbonate, Ca3) anhydrite (Deckanhydrite or upper Staßfurt anhydrite, A2r; Haupt or Leine anhydrite, A3; Pegmatite or Aller anhydrite, A4), and potash salts (Staßfurt potash seam, K2; Ronnenberg potash seam, K3Ro; local occurrences of the Bergmannssegen potash seam, K2Be; Riedel potash seam, K3Ri). After BORNEMANN et al. (1988) the thicknesses for the various evaporites in the Gorleben salt dome are estimated to be 40 m of clay, 80 m of anhydrite, 935-970 m of rock salt, and 20-25 m of potash salt (carnallitite).

Information on the compositional changes which occurred in the geological past is to be found above all in the K-Mg mineral associations of the three potash seams. At temperatures below 100 °C potash salts react sensitively to temperature changes and to the influence of unsaturated aqueous solutions which cause certain changes in the mineralogical and chemical composition of the original evaporite deposits. This is especially true for occurrences of $MgSO_4$-bearing minerals, in combination with Na-, K-, and/or Mg-chlorides (Staßfurt potash seam). Neither clay, limestone, dolomite, gypsum-anhydrite, nor rock salt are significant indicators of mineral reactions and material transport like potash salts. Therefore, potash salt seams, particularly the Staßfurt salt seam, are of great importance in the evaluation of the mineralogical and chemical evolution of the 250-million-years-old rocks of the Gorleben salt dome and of other Zechstein evaporites in the Hannover region.

The dominant evaporites of the Gorleben salt dome have the following mineralogical composition (BORNEMANN et al. 1988):

## *Haupt or Leine anhydrite, A3*

(based on analyses of GUNDLACH, Bundesanstalt für Geowissenschaften und Roh-stoffe)

| Mineral | Mass fraction in % |
|---|---|
| anhydrite | 93.0 |
| magnesite | 5.0 |
| clay, carnallite | |
| other minerals | 2.0 |

## *Leine rock salt, Na3 (Banksalz, Bändersalz intercalated between the K3Ro and K3Be or K3Ri potash seams)*

(compiled by BORNEMANN et al. 1988, based on analyses of MÜLLER-SCHMITZ 1985)

| Mineral | Mass fraction in % |
|---|---|
| halite | 97.1 |
| polyhalite | 2.0 |
| carnallite | 0.5 |
| anhydrite | 0.4 |
| water-insoluble minerals (predominatly clay) | ≤ 1.0 |

## *Leine rock salt, Na3 (Basissalz, Liniensalz, Orangesalz below the K3Ro potash seam)*

(compiled by BORNEMANN et al. 1988, based on analyses of MÜLLER-SCHMITZ 1985)

| Mineral | Mass fraction in % |
|---|---|
| halite | 94.4 |
| anhydrite | 5.0 |
| polyhalite | 0.5 |
| carnallite | < 0.1 |
| water-insoluble minerals (predominantly clay) | ≤ 1.0 |

## *Staßfurt rock salt, Na2*

(compiled by BORNEMANN et al. 1988, based on analyses of MÜLLER-SCHMITZ 1985)

| Mineral | Mass fraction in % |
|---|---|
| halite | 95.0 |
| anhydrite | 4.9 |
| polyhalite | 0.1 |
| water-insoluble minerals (predominantly clay) | ≤ 1.0 |

## Riedel potash seam, K3Ri

(based on data in Bornemann et al. 1988)

Reddish-brown to dark-gray, fine- to medium-grained rock salt. There are flakes and clouds of dark-gray clay and anhydrite in the rock salt. Red and yellow interstitial carnallite is distributed irregularly through the matrix of the rock salt.

## Bergmannssegen potash seam, K3Be

(based on data in Bornemann et al. 1988)

The potash seam found in the laminated salt in exploratory drilling Gorleben 1003 is equivalent to the K3Be, but not to K3Ro. The K3Be is composed of brecciated carnallitite (Trümmercarnallit), comparable with that of the Staßfurt potash seam. In the lower portion of exploratory drilling Gorleben 1005 the K3Be potash seam has been identified as a rock salt horizon. Due to limited reconnaissance the overall distribution of K3Be in the Gorleben salt dome is not yet known.

## Ronnenberg potash seam, K3Ro

(based on data in Bornemann et al. 1988)

One horizon corresponding to the Ronnenberg potash seam was encountered in shaft drilling 5001. It consisted predominantly of halite containing kieserite, langbeinite, polyhalite, anhydrite, and small amounts of sylvite usually as laminations and bands. Fischbeck (1984) interprets this mineral association to be an impoverishment of K3Ro.

## Staßfurt potash seam, K2

(based on data in Bornemann et al. 1988, analyses of Gundlach, Bundesanstalt für Geowissenschaften und Rohstoffe)

The Staßfurt potash seam was encountered in the form of brecciated carnallitite in all four deep drillings (i.e., exploratory drillings Gorleben 1002-1005 and shaft drilling 5001). Layers and flakes of white to dark-gray kieserite and subordinate anhydrite as well as fragments of rock salt (light-gray, fine-crystalline rock) are embedded in a matrix of red, violet, white, or clear carnallite. The Staßfurt potash seam has an average thickness of 20-30 m. There are folds with amplitudes up to 70 m in K2, as evidence of its tectonic history. Hydrocarbonous gases have been detected in the carnallitite (crackle carnallitite) in nearly all outcrops of K2.

| Mineral | Mass fraction in % |
|---|---|
| halite | 57 |
| carnallite | 25 |
| kieserite | 16 |
| anhydrite | 1 |
| water-insoluble minerals (predominantly borates and clays) | ≤ 1 |

The salt wash surface of and the Quaternary channel in the Staßfurt potash seam were studied with core material from salt table drillings GoHy 1301-1305, south of drilling GoHy 1141. The five boreholes were sunk in a row and spaced 15 to 80 m apart (Fig. 27). The drill sites are located about 600 m northeast of exploratory drilling Gorleben 1002 on the overturned NE flank of the salt dome (BORNEMANN et al. 1988). Studies have shown that the K- and Mg-bearing minerals of K2 about 90 - 130 m below the salt table (about 290 m below the earth's surface) have been completely dissolved (zone of impoverishment). The original mineral association was replaced by a horizon of red rock salt which contains voluminous infillings of clastic material from the overlying rock. BORNEMANN et al. (1988) point out that the original stratigraphic relationships of K2 can still be reconstructed in spite of the intense fracturing and infillings.

If the Staßfurt potash seam originally consisted of kieseritic carnallitite, a kainite rock is expected to have been formed due to the influence of unsaturated aqueous solutions below 72 °C on the initial carnallitite. In fact, the K2 potash seam was encountered in the form of a kainitic rock in drilling GoHy 1304.

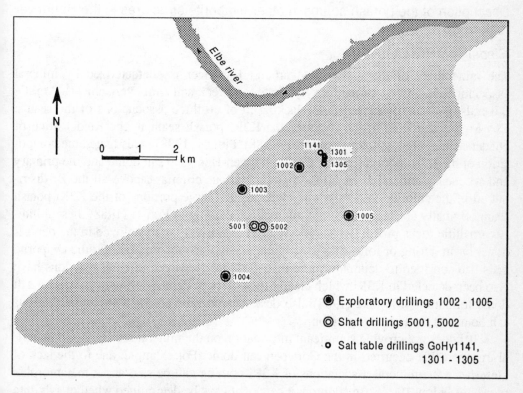

**Fig. 27** Map showing the location of exploratory drillings, shaft drillings, and several salt table drillings in the Gorleben salt dome, Germany.

Formation water (groundwater) from the hanging-wall rocks has obviously affected the Staßfurt potash seam to differing extent at various depths. BORNEMANN et al. (1988) pointed out that in drilling GoHy 1305, K2 was encountered as brecciated carnallitite, and impoverishments and kainite were absent.

In the area around boreholes GoHy 1304 and 1305 the lower boundary of the influence of unsaturated aqueous solutions from the hanging-wall rocks on the Staßfurt potash seam is apparently 140 - 170 m below the salt wash surface (about 445 m below the earth's surface). Tests in both boreholes have demonstrated that the salt is impermeable, with the exception of the uppermost sections (BORNEMANN et al. 1988).

The effect of solutions on the Staßfurt potash seam must have been limited to the Elster glaciation (about 300 000 - 600 000 years ago) as evidenced by the intercalations of clastic material from the overlying rocks in the evaporites below the salt wash surface and the formation of caprock over the Gorleben salt dome. In the stratigraphic sequence of the GoHy 1141 and 1301 - 1304 drillings, however, there was no evidence that significant amounts of evaporite rocks were dissolved following the Elster glaciation. Evidence of post-Elster subrosion was only found in drilling GoHy 1305 (BORNEMANN et al. 1988).

## Description of the potash seams in other evaporite structures in the Hannover region

### Riedel potash seam, K3Ri

The mineralogy of the K3Ri seam around Hannover is characterized by mineral associations including carnallitite, sylvinite, and rock salt (impoverishment). $MgSO_4$ minerals, particularly kieserite, are lacking in of all three associations of the seams. The $MgSO_4$ contents in the rocks of the K3Ri potash seam in the Riedel mine of Niedersachsen are less than 1% (PETERS 1988). PETERS (1988) investigated the composition of the K3Ri potash seam in the Wathlingen-Hänigsen salt dome and the primary and secondary mineral associations therein. Various criteria - above all the Br distribution in the chloride minerals - are evidence that large portions of the K3Ri potash seam originally consisted of carnallitic rock. According to PETERS (1988) these initially carnallitic rocks probably had already been extensively recrystallized in the depositional basin during or following their initial crystallization and not after the evaporite beds had subsided to depths of more than 1000 m. However, mineral reactions have also been detected in K3Ri which clearly took place at depth (during or following salt dome formation?). PETERS (1988) also describes occurrences of K3Ri in the various salt domes in the Hannover region.

K3Ri does not provide significant information on the mineral reactions and material transport that occurred in the Gorleben salt dome. For example, due to the lack of kieserite in the mineral associations of K3Ri nothing can be said about metamorphic processes below 100 °C. In addition, it cannot always be determined whether sylvinite (a halite-sylvite-bearing rock) formed from a carnallitite under the influences of solutions at the surface or after subsidence of the evaporites to greater depths.

## *Ronnenberg potash seam, K3Ro*

In some salt structures of the Hannover area carnallitic rocks, sylvinite, and rock salt (impoverishments) have been observed in the Ronnenberg potash seam. A whitish-gray, layered sylvinite is the most dominant rock in K3Ro.

Although the kieserite contents average ± 3% of K3Ro and are thus somewhat higher compared with K3Ri, they are still substantially less than those of the Staßfurt potash seam containing 14-18% (HERRMANN 1991c).

The K3Ro seam in the Gorleben salt dome is obviously of no value for evaluating mineral reactions and material transport, as is true for K3Ri. However, this can change if new outcrops of these potash seams are exposed during underground exploration.

Due to the different amounts of the $MgSO_4$ minerals kieserite and kainite the widespread occurrences of Z3 potash seams in the Hannover region are described as ranging from sulfate-type (e.g., Staßfurt seam) via an intermediate-type (i.e., K3Ro) to chloride-type (K3Ri) marine evaporites (HERRMANN 1991c).

The occurrences of the Ronnenberg potash seam in the various salt structures in the Hannover region are described, for example, by PETERS (1988).

## *Staßfurt potash seam, K2*

The Staßfurt potash seam is found in all salt domes in the Hannover region. The K2 evaporites are brecciated carnallitite, kieseritic Hartsalz, and K-free rock salt (impoverishments). The essential difference between K2 and the K3Ro and K3Ri potash seams lies in the comparatively high kieserite contents of the carnallitite and Hartsalz (sulfate type, see Ronnenberg potash seam). Kieseritic Hartsalz (halite-sylvite-kieserite) has not been found in any of the K2 outcrops in the Gorleben salt dome. However, the K2 potash seam could still be encountered as Hartsalz in the course of further exploration in the Gorleben salt dome, because the high $MgCl_2$ contents in the solutions found in core material from exploratory drillings Go 1002-1005 must originate from the dissolution and decomposition of the mineral carnallite. To date, however, larger amounts of carnallite have only been found in the carnallitic rock of the K2 potash seam. PETERS (1988) provides information on the previous studies of the formation and distribution of the K2 potash seam in the Hannover region.

Of the four potash seams in the Zechstein evaporites of the Gorleben salt dome, only the mineral associations of the Staßfurt potash seam are of great importance to all questions connected with mineral reactions and material transport.

## 17.2  Fracture fillings

In addition to mineral associations of potash seams, secondary minerals occurring as crack and fissure fillings in evaporites also provide important information on the material transport which occurred in an evaporite sequence following deposition in a sedimentary basin.

Cracks and fissures, which for the most part have been healed with secondary minerals, have been observed in all previously mined salt domes of the Hannover region (e.g., HERRMANN 1983a:157ff). The Gorleben salt dome is no exception, as evaluations of core material from exploration drillings, shaft drillings, and salt table drillings have shown. Tab. 12 shows an overview of the crack and fissure fillings (secondary minerals) in the rocks of the Gorleben salt dome based on the studies of BORNEMANN et al. (1988), FISCHBECK & BORNEMANN (1988), and BORNEMANN (1991).

Of the secondary fracture-filling minerals, the chloride minerals dominate, halite being the most common. In exceptional cases even bischofite has been found as fracture filling, e.g., in the Asse salt dome southeast of Wolfenbüttel (e.g., RICHTER & KLARR 1984).

The fracture-filling minerals provide important information on the composition of the salt solutions from which they crystallized. There are many indications that Mg-

**Tab. 12** Fractures and fracture-filling minerals in the evaporite rocks of the Gorleben salt dome (BORNEMANN et al. 1988; FISCHBECK & BORNEMANN 1988; BORNEMANN 1991). n.d., no data given.

| Rock type | Stratigraphy | Number of fractures | Dimension of fractures [m] | Depth below surface [m] | Fracture-filling minerals |
|---|---|---|---|---|---|
| Clay | lower Aller-clay, T4 (Roter Salzton) | numerous fractures | n.d. | n.d. | fibrous halite |
| | Leine-clay, T3 (Grauer Salzton) | intensely fractured | n.d. | n.d. | fibrous halite, fibrous carnallite |
| Anhydrite | Leine-anhydrite, A3 (Haupt-anhydrite) | occasional fractures | n.d. | n.d. | carnallite, sylvite, halite, occasional borate coatings on fracture surface |
| Rock salt | Aller-rock salt, Na4 (Schneesalz, Rosensalz) | occasional and irregularly distributed mm- to 2 cm-size carnallite clusters (secondary?) in the rock salt beds, no fractures | | | |
| | Leine-rock salt, Na3 (Na3tm, Tonmittelsalz) | frequent fractures | n.d. | n.d. | fibrous halite, fibrous carnallite |
| | Leine-rock salt, Na3β (Liniensalz) | subordinate fracturing | n.d. | n.d. | carnallite, but not in fractures |
| | Staßfurt-rock salt, Na2 (Hangend-salz) | occasional fractures in drilling Go 1005 | > 0.1 | 421.7 | halite, crystals > 1 cm in diameter |

free minerals such as halite and sylvite also crystallized from $MgCl_2$-bearing solutions (see Chapter 17.3, solutions).

With decreasing temperature these minerals can crystallize from solutions of uniform composition (model for polythermal mineral formation) and/or mixtures of various solutions of differing chemical compositon (e.g., NaCl-saturated solutions mix with $MgCl_2$-concentrated solutions). The second model is applicable to both polythermal and isothermal conditions. In both cases it must be presumed that salt solutions from other areas of the evaporite body have invaded the voids of the cracks and fissures, followed by crystallization of the secondary minerals. It is evident that the salt solutions originated from other parts of an evaporite sequence - of differing petrography - when the chemical compositions of the secondary minerals differ from that of the rocks in which the fractures and secondary minerals have formed. Examples of this are all compounds which crystallized in the fractures in anhydrite beds, with the exception of $CaSO_4$. The same is also true for carnallite and halite with high Br contents in fractures which pass through rock salt horizons (see, e.g., FISCHBECK & BORNEMANN 1988). The fracture-filling secondary minerals mentioned above are normally fine to coarse grained.

One exceptional feature of some fracture-filling secondary minerals is their fibrous form (e.g., the fibrous halite and carnallite-filling fractures in the clay and rock salt of the Gorleben salt dome). The formation of the fibrous minerals cannot be explained conclusively in all cases. The explanation given by SCHMIDT (1911, 1914) and MÜGGE (1928) that the fibers grew as pore water from the host rock penetrated into the fractures as they opened is probably true for the clays. By determining the Br contents of different parts of such fibers HERRMANN (1964) was able to provide support for the formation of fibrous halite in a fracture in a clay-marl bed (Marie-Louise mine in Alsace, Oligocene salt deposits in the upper Rhein basin) by the mechanism postulated by SCHMIDT (1911, 1914) and MÜGGE (1928). Is this also true for fibrous fracture-filling minerals in chlorides such as rock salt (e.g., Na3 in Gorleben)?

Although no pore solutions in the host rock were necessary for the formation of the fine- to coarse-crystalline secondary minerals in fractures, the host rock must contain solution-filled pore space for the formation of fibrous minerals in the model of SCHMIDT (1911, 1914) and MÜGGE (1928).

## 17.3 Solutions

Solutions of differing chemical composition and concentration have been found in all surface and underground exposures of all evaporite bodies explored in the Hannover region. They are trapped in evaporites and escape through various paths when the host rocks are exposed during drilling or mining. Solutions have been observed in all known evaporite occurrences of the world. Hence, the fluid constituents of an evaporite body are just as essential to the composition as the solid minerals and rocks. However, clear distinctions are to be made between solutions which are genetically related to the rock and those which penetrate into the rock during exploration and mining. In this context it is useful to divide salt solutions into three groups (Tab. 13).

Only the solutions of group 1 (geology, deposits) are to be considered for studies of the evolution of evaporite composition. Solutions of group 2 (mining) can form during the exploration and operation of a repository mine, but are of no importance to questions concerning the long-term safety of repositories in evaporites. Solutions of group 3 are not applicable to a repository mine, but group 3.2 is to be considered for the case of backfill.

**Tab. 13**  Origin of salt solutions in potash and rock salt mines (from HERRMANN 1983a: 174).

| Group 1: | Geology, deposit | Group 2: | Mine | Group 3: | Factory |
|---|---|---|---|---|---|
| 1.1 | surface and groundwater | 2.1 | shaft solutions | 3.1 | hydraulic packing |
| 1.2 | formation water | 2.2 | ventilation solutions | 3.2 | dry packing |
| 1.3 | metamorphic solutions | 2.3 | drilling solutions | 3.3 | solutions from crude-salt processing |
| 1.4 | residual solutions from the depositional basin | 2.4 | shaft sump | | |

The solutions of group 1 occurring in marine evaporites can be described with the following systems for various temperatures and utilized in calculating mineral reactions:

$NaCl - H_2O$

$KCl - H_2O$

$MgCl_2 - H_2O$

$CaCl_2 - H_2O$

$NaCl - KCl - H_2O$

$NaCl - MgCl_2 - H_2O$

$NaCl - CaCl_2 - H_2O$

$NaCl - CaSO_4 - H_2O$

$MgCl_2 - CaCl_2 - H_2O$

$NaCl - KCl - MgCl_2 - H_2O$

$NaCl - KCl - MgCl_2 - CaCl_2 - H_2O$

$2NaCl + MgSO_4 \Leftrightarrow Na_2SO_4 + MgCl_2$

$NaCl - KCl - MgCl_2 - MgSO_4 - H_2O$

The chemical composition and concentration of the solutions obtained from exploratory drillings Go 1002-1005 are listed in Tab. 14 and Fig. 28. All these solutions have high $MgCl_2$ contents (290-440 g $MgCl_2$/l). Solutions with high $MgCl_2$ contents have been encountered again and again over the many decades of potash and rock salt mining in central and northern Germany. Among the scientifically interesting questions is that of the formation of bischofite-saturated solutions concurrent with the lack of bischofite as a mineral or rock in the evaporite sequence (e.g., BRAITSCH 1962: 100). It is also noteworthy that $CaCl_2$ occurs in the solutions of three out of four drillings while $MgSO_4$ was contained in the solution of only one drilling, and consequently $CaCl_2$ was lacking. The chemical composition of the solutions stored in anhydrite and rock salt evidences that the fracture-filling minerals found in clay, anhydrite, and rock salt obviously crystallized from solutions with high $MgCl_2$ contents (see HERRMANN & v. BORSTEL 1991; v. BORSTEL 1992).

There is a multitude of information and analytical data on the occurrence and composition of salt solutions from salt structures mined in the Hannover region over the past 60-70 years. The concentration and composition of many of these solutions are comparable with the solutions encountered in the Gorleben salt dome. Unfortunately, only a fraction of the observations and data from the area of potash and rock salt mining is available for purposes of scientific research. The vast majority of information is to be found in unpublished reports in the archives of mining bureaus and mining companies and are extremely difficult to gain access to. Hence, 60 years (!) after its publication the study of BAUMERT (1928, modified 1952) is still an important (since generally accessible) source of information on the occurrence of solutions in Zechstein evaporites of central and northern Germany. However, when reading the study of BAUMERT (1928) it must be remembered that in past decades salt solutions in potash and rock salt mines were not sampled with the necessary care. When sampling solutions from very slowly dripping sources for hours or days, a considerable portion of the water contained in the solution can easily evaporate, resulting in a change in the original concentration and chemical composition of the salt solutions (e.g., HERRMANN 1982). In this way the determination of the evolution and origin of the solutions released by the evaporites is made more difficult, if not impossible.

Hence, a study incorporating current knowledge on the occurrence, composition, and evolution of salt solutions in potash and rock salt mines is one of the most urgently needed for evaluating the safety of underground repositories for anthropogenic wastes (whether radioactive or nonradioactive) in horizontal and steeply inclined salt strata. Such a study is currently being prepared at the Institute of Mineralogy and

**Tab. 14** Composition and concentration of solutions from exploratory drillings Go 1002-1005 (from HERRMANN 1983b: 39).

| Drilling | Compounds | Concentration [g/l] | Mass fraction [%] | mol/1000 mol $H_2O$ |
|---|---|---|---|---|
| Gorleben 1002 | NaCl | 24.7 | 1.93 | 8.47 |
| | KCl | 28.6 | 2.23 | 7.67 |
| | $MgCl_2$ | 307 | 24.0 | 64.6 |
| | $CaCl_2$ | 22.1 | 1.73 | 4.0 |
| | $CaSO_4$ | 0.44 | 0.03 | 0.06 |
| | $CaCO_3$ | 0.49 | 0.04 | 0.1 |
| | $H_2O$ | 897 | 70.0 | 1000 |
| | Br | 4.10 | 0.321 | |
| | Li | 0.042 | 0.0034 | |
| Gorleben 1003 | NaCl | 12.7 | 0.97 | 4.3 |
| | KCl | 6.7 | 0.51 | 1.8 |
| | $MgCl_2$ | 384 | 29.2 | 80.2 |
| | $MgSO_4$ | 10.1 | 0.77 | 1.7 |
| | $CaSO_4$ | 2.4 | 0.18 | 0.35 |
| | $H_2O$ | 896 | 68.4 | 1000 |
| | Br | 7.02 | 0.535 | |
| | Li | 0.066 | 0.0050 | |
| Gorleben 1004 | NaCl | 40.8 | 3.2 | 14.0 |
| | KCl | 35.9 | 2.8 | 9.7 |
| | $MgCl_2$ | 287 | 22.7 | 60.5 |
| | $CaCl_2$ | 3.5 | 0.28 | 0.63 |
| | $CaSO_4$ | 2.2 | 0.17 | 0.32 |
| | $H_2O$ | 894 | 70.9 | 1000 |
| | Br | 2.17 | 0.172 | |
| | Li | 0.023 | 0.0018 | |
| Gorleben 1005 | NaCl | 4.13 | 0.31 | 1.44 |
| | KCl | 1.42 | 0.10 | 0.39 |
| | $MgCl_2$ | 439 | 32.4 | 94.1 |
| | $CaCl_2$ | 29.9 | 2.21 | 5.5 |
| | $CaSO_4$ | 0.07 | 0.01 | 0.01 |
| | $H_2O$ | 879 | 65.0 | 1000 |
| | Br | 7.26 | 0.537 | |
| | Li | 0.193 | 0.0143 | |

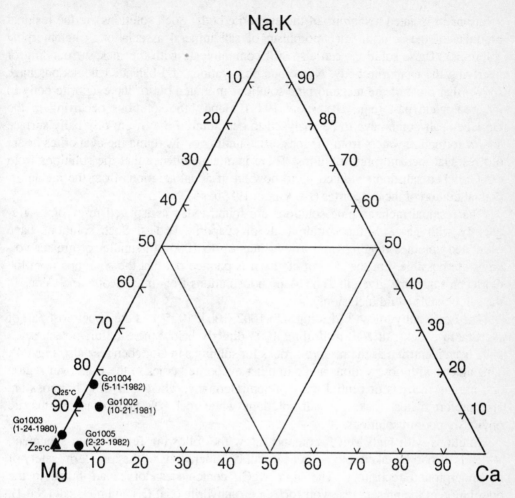

**Fig. 28** Equivalent percentages of the cations of solutions from drillings Go 1002-1005 in the Gorleben salt dome. Solutions $Z_{25°C}$ and $Q_{25°C}$ are also plotted.

Mineral Resources, Dept. of Salt Deposits and Underground Repositories, Technical University of Clausthal, Germany (v. BORSTEL 1992).

The surface waters, groundwaters, and formation waters of group 1 in Tab. 13 are solutions occurring in the rock overlying and/or surrounding evaporite bodies. When such waters or unsaturated solutions penetrate into salt deposits, they react primarily with very soluble minerals such as halite, sylvite, and carnallite, in the course of which, solutions are formed whose composition differs from that of the waters surrounding the evaporite body. Based on the chemical composition of the solutions it can normally be inferred whether or not the solutions are residues of geologically old solutions in isolated reservoirs of the evaporite body. Such solutions are the residual products of the solution metamorphism of salt mineral associations (metamorphic solutions). These solutions can also still communicate with the rock surrounding or overlying the evaporite body. Numerous observations and balance calculations have shown that most of the metamorphic solutions migrated out of the evaporite body in the geological past (e.g., HERRMANN 1979). Hence, the solutions occurring in the Gorleben salt dome have to be analyzed to determine if they were originally surface and/or formation waters from the rocks surrounding or overlying the evaporites or are old residual metamorphic solutions. There is much evidence that the solutions from the Gorleben salt dome studied up to now are residual solutions from the metamorphic alteration of the evaporites (HERRMANN 1983b: 45).

The residual metamorphic solutions are found today as isolated reservoirs - frequently with gas - within certain beds of evaporite bodies. Such solutions have estimated volumes ranging from a few liters up to 1000 m³; smaller or greater volumes are possible. For the sake of clarity it is pointed out that the solutions from the Gorleben salt dome given in Tab. 14 are not solutions from microscopic inclusions in the salt crystals (fluid inclusions).

The Leine anhydrite A3 (drillings Go 1002, 1004, 1005) and the uppermost part of the Staßfurt rock salt Na2 in drilling 1003 directly below the Staßfurt potash seam have been identified as the reservoir rocks for solutions in Gorleben (see also Fig. 24). This agrees with observations made in other evaporite bodies in the Hannover region and mining districts of central and northern Germany, where anhydrite horizons are also preferred host rocks for salt solutions while rock salt and potash seams are obviously poor in solutions.

Solutions with high $MgCl_2$ contents allow the following fundamental statement, which is also to be taken into account for all considerations of the long-term safety of an underground repository: The high $MgCl_2$ contents cannot be attributed to the dissolution of the present reservoir rocks, i.e., anhydrite ($CaSO_4$) and rock salt (NaCl).

Based on current knowledge there is only one rock within the Gorleben salt dome which was able to contribute the $MgCl_2$ component to the solutions from exploratory drillings Go 1002-1005. This rock is carnallitite which in the Gorleben salt dome occurs primarily in the Staßfurt potash seam and locally in the Bergmannssegen potash seam of the Leine sequence (see Chapter 17.1). The $MgCl_2$ could originate from the mineral bischofite ($MgCl_2 \cdot 6\ H_2O$), but bischofite has not yet been identified in the Gorleben salt dome. According to recent investigations on fluid inclusions from the Leine rock salt in the Gorleben salt dome the high $MgCl_2$ concentrations are residues of the dissolution of bischofite by seawater. The bischofite has been crystallized above the Staßfurt potash seam (K2) in the Zechstein basin and dissolved during the Zechstein 3 (HERRMANN & v. BORSTEL 1991).

The $CaCl_2$ constituent in the salt solutions can have the following sources (e.g., KOCKERT 1969): (1) dissolution of $CaCl_2$-bearing minerals such as tachhydrite ($CaMg_2Cl_6 \cdot 12\ H_2O$), (2) reaction between $MgCl_2$-bearing solutions and carbonates of the salt clay, (3) organic reduction of sulfate with the evaporite rocks, and (4) the ingression of $CaCl_2$-bearing formation waters from the surrounding rock into the salt dome at an unknown time in the geological past.

The salt solutions encountered in the Gorleben salt dome also confirm another fact recognized in all other evaporites: The present reservoir rock of many salt solutions is not the site or rock in which the solutions formed. Since the horizons from which the $MgCl_2$ contents of the solution originated were spatially separated from the present reservoir rock, the solutions must have migrated within the evaporite rocks from the source rock to the reservoir rock over undeterminable distances (Chapter 16, Fig. 25). Not only anhydrite and clay horizons but also rock salt can host the paths for solution migration. This conclusion is based on the observation of fracture-filling minerals in the clay, anhydrite, and rock salt of the Gorleben salt dome (Chapter 17.2, Tab. 12).

The migration of solutions into salt structures of northern Germany is not limited to depths of less than 1000 m. In drillings 1002 and 1005 of the Gorleben salt dome solutions were also released from the Leine anhydrite and the surrounding rock between 1000 and 1900 m depth. This finding is significant to the interpretation of the evolution of solution reservoirs and the flow of salt solutions in the Gorleben salt dome, and consequently to the assessment of the long-term safety of an underground repository.

Based on the current status of exploration of the Gorleben salt dome and considering the experience with other salt domes in the Hannover region and evaporites in central and northern Germany there probably are solution reservoirs in limited areas of the Gorleben evaporites (anhydrite, more rarely rock salt). Hence, such reservoirs are just as likely to be encountered during the mining exploration of the Gorleben salt dome as in other, horizontal and steeply inclined salt bodies. It must also be remembered that after solution has flowed out of an initially limited reservoir, paths to the surrounding or overlying rock are reopened, which could cause the release of even more solution.

The potash and salt mining activities in central and northern Germany demonstrate that solutions in evaporites do not make underground mining impossible. This conclusion which is based on 120 years of experience is equally true - under certain conditions - for the construction of underground repositories for anthropogenic wastes in evaporites. In other words, limited solution reservoirs in a salt dome isolated from the surrounding and overlying rock are not necessarily an argument against the construction of underground repositories in such a body of rock. Today, the problem presented by salt solutions can usually be recognized in time, evaluated, and dealt with during the operating phase of a salt mine based on past mining experience. Thoughts on how this experience can be applied to questions regarding the long-term safety of underground repositories will be discussed in Chapter 19.

## 17.4 Gases

In addition to the solids and liquids in marine evaporite deposits, there are also gases trapped in the minerals and rocks. Previous observations have confirmed again and again that the marine evaporites occurring in differing geological formations nearly always consist of solid, liquid, and gaseous compounds and, in part, elements. HERR-MANN (1988a) presented a discussion and summary of current knowledge on gases in marine evaporites in the context of repositories for radioactive and nonradioactive wastes in evaporites (see also Part II). This study also considers the published information on the occurrence of gases in salt structures in the Hannover region, especially in the Gorleben salt dome (see also Chapter 14).

Unfortunately, there are hardly any available data on the occurrence of salt-bound gases from the potash and rock salt mining industry, as is true regarding solutions as well. There are more data on the gas occurrences in unpublished reports of mining bureaus and the mining industry than in the generally accessible scientific literature. This is also true for the Zechstein occurrences in Hessen and Lower Saxony. In contrast to the situation in the Federal Republic of Germany in the borders prior to 1990, a scientific research project on salt-bound gases was jointly conducted by various institutions from differing areas of science in the 1950s and 1960s in the former German Democratic Republic. The majority of the analytical results were published in the scientific literature and are consequently available to research.

The following observations for gases in the Gorleben salt dome are pointed out again for the purposes of this chapter: Gases and fluid condensates were released from the Leine rock salt (Na3) in shaft drillings Go 5001 and 5002 in the center of the salt dome. The gases were trapped in minerals and fractures. For example, Orange salt (Na3) occurred in the form of crackle salt in Go 5001 at a depth of 966.55 m (GRÜBLER & REPPERT 1983: 49). AKSTINAT (1983) yielded the following compositions for gas samples from Go 5002, which were obviously not contaminated with air: 55-88 vol% $N_2$, 10-36 vol% $CH_4$, 0.05-0.1 vol% $CO_2$, some higher hydrocarbons. In a gas mixture of hydrocarbons, methane dominated with about 82 %, followed by 9 % ethane, 5 % propane, 2 % butane, and < 1.5 % other hydrocarbons. Hydrocarbon-rich gases and $N_2$-$CO_2$ mixtures presumably originated from releases of greatly differing intensity (AKSTINAT 1983). Previous studies indicate that the gases originated from silicate rocks of the basal Zechstein, e.g., Kupferschiefer, T1 (e.g., BORNEMANN et al. 1988; GERLING et al. 1991).

The carnallitic rock of the Staßfurt potash seam occasionally occurs as crackle salt (crackle carnallite) and was encountered at various depths in core material from the exploratory drillings (e.g., BORNEMANN et al. 1988). The gas in this rock consisted of a hydrocarbon-bearing mixture making up about 1500 µg/kg of rock (ppb). The layers and clusters of carnallite found in core material from the exploratory drillings in the Leine rock salt (Na3β, Liniensalz) crackled and released hydrocarbonous gases when struck (BORNEMANN et al., 1988).

The occurrence of crackle salt is of interest to this study especially in the following respect: Mineral-bound gases (e.g., in crackle salts) can be fixed in minerals in the presence of aqueous solutions during mineral reactions and recrystallization. Hence, the occurrence of crackle salt is evidence of the fact that minerals were altered and/or recrystallized at one time in this part of the evaporite body. Therefore, the following criteria must be considered with respect to the occurrence of crackle salt in the Leine rock salt (Na3) of the Gorleben salt dome:

i.  Is the gas-bearing crackle salt particularly coarse crystalline, colorless, and transparent or brownish (condensate)?

ii. How much Br does the rock salt contain? Are there any indications of recrystallization in Na3β?

Similar studies and considerations must be made for the crackle carnallite in the Staßfurt potash seam and the Leine rock salt (Na3β).

Gases occur frequently with solutions (e.g., HERRMANN 1988d: 20). This also applies to salt domes in Niedersachsen. BRAITSCH (1962: 207; 1971: 267) pointed out that the gases occurring together with solutions are probably an important source of energy for solution migration. In such cases the escaping quantities of solutions behave in accordance with the Boyle-Mariotte law; the time dependency must be considered to be a measure of the decrease in pressure, which can be attributed to gas expansion.

Consequently, the effectiveness of the gases fixed in the salt as possible driving force for solutions is of particular interest with respect to the long-term safety of underground repositories.

# 18 Calculation of mineral reactions and material transports

## 18.1 Fundamentals for data processing

### Concept

Observations in nature such as the geological setting and the mineralogical and chemical composition of an evaporite body serve as the basis for scientific analysis (see also Fig. 26). More and more data will become available as underground exploration of a salt dome progresses (e.g., Gorleben). A computer program for processing the available data must have the following features:
- flexibility,
- user-friendliness,
- flawless operation,
- sufficient capacity,
- speed.

The following data processing functions (subroutines) must be linked for evaluating the data:
- acquisition and storage especially of data on the geological setting and on solid, liquid, and gaseous composition of the evaporite body in a relational data base with possibilities for inquiry according to the most varied keywords.
- calculation of various characteristic values for salt solutions (e.g., mass fraction in %, g/l, mol/1000 mol $H_2O$, equivalent percentages of anions and cations),
- calculation of mineral reactions,
- calculation of volumes and masses of the evaporites and solutions that participated in the reactions (material balance calculations),
- computerized evaluation of calculations with graphic presentation of the results.

A modular program system with a common user interface was developed on standard personal computers to convert the theoretical model for actual application. Several of the individual program modules, referred to as subroutines in the following, use standard software. In this way maximum efficiency and flexibility for every module (subroutine) was able to be attained.

For our purposes the computer capacity was also modularly structured to be able to process the amount of data expected in the course of the underground exploration of a salt dome. This is done by networking several PC's and means that several data stations each with local intelligence can be employed independently for data acquisition and processing. In this way the data bases are always kept up to date, even with great amounts of incoming data. However, only efficient data processing can assure reliable information for evaluating long-term safety. There are also hierarchial access authorizations which assure the necessary data security.

The program system described in more detail in the following can process data for evaluating long-term safety, independent of the repository site. However, it must be pointed out again that thorough knowledge of the stable and metastable solution equilibria of marine evaporites is indispensable even when using this system. The system acts solely as a »tool for thought« for the subsequent interpretation of the evolution of an evaporite body in the geological past and the attempt to prognose its future development.

## Data bases for evaporites and salt solutions

Two data bases were set up for storage and statistical evaluation. In the EVAPO-RITES data base, complete rock analyses with data on mineralogical and chemical constituents are to be stored.

In the SALT SOLUTIONS data base, all data on the results of the subroutine CHARACTERISTIC VALUES OF SALT SOLUTIONS are to be stored (see below). Subroutines allow statistical evaluations. A nearly arbitrary number of codes and keywords can be used for data inquiries: for example, »list of all solution analyses before 1-1-1970 from the Staßfurt rock salt with < 25 % $MgCl_2$«. A graphic software is linked with the data base via dynamic data exchange (DDE) so that the listed solutions can be entered in a diagram with the equivalent percentages for anions and cations. The generation of plots with a standard graphic program has the advantage that the graphics can be changed easily later: for example, $Z_{25°C}$ and $Q_{25°C}$ in Fig. 28 were added later.

## Calculation of the characteristic values of solution analyses

The program part CHARACTERISTIC VALUES OF SALT SOLUTIONS is struc-tured based on the program published by HERRMANN et al. (1978), which has been improved with several possibilities for entry and conversions (Fig. 29). The calcula-tion for a solution from the Gorleben exploratory drilling Go 1002 is given in Fig. 30 as an example. The calculations can be used not only for statistical evaluation, for example, but also for calculations in the subroutine MINERAL REACTIONS (see below).

## Calculations of mineral reactions

Using microscopy, X-ray, and chemical methods it can be determined whether the composition of the rocks in a salt dome is primary or secondary. In other words it has to be determined whether the composition of the evaporites which precipitated from seawater has been altered partially or completely by aqueous solutions and/or tempe-

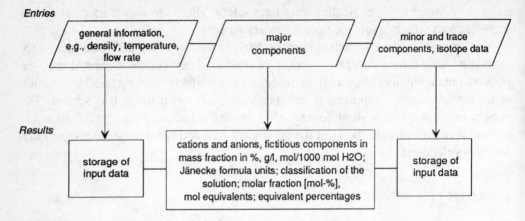

Fig. 29 Structure of the subroutine CHARACTERISTIC VALUES OF SOLUTION ANALYSES.

rature. For example, did the Staßfurt potash seam originally consisted of primary carnallitite or of rocks such as kieseritic or anhydritic Hartsalz, sylvinite, or impoverishments?

If the original mineral association halite + carnallite + kieserite (carnallitite; genetically classified based on Br contents) is dominant in the Staßfurt potash salt seam, unsaturated aqueous solutions were no longer active at least at the level of the potash salt rocks since deposition of the evaporites in the Zechstein basin 250 Ma ago.

In contrast, if for example the mineral association halite + kainite + kieserite is widespread in the Staßfurt potash salt seam, the mineralogical and chemical composition of an originally carnallitic rock must have been changed by unsaturated aqueous solutions at temperatures < 72°C (solution metamorphism). In this case the observed or inferred composition of the neogenic kainitic rock can be used to quantify solutions which participated in the metamorphism of certain masses of evaporite rocks (Chapters 18.2.1, 18.2.2).

The subroutine MINERAL REACTIONS was written to quantify this metamorphism. It allows the isothermal and polythermal calculation of solution-metamorphic processes based on the solution equilibria of marine evaporites. This program was structured after the program LOESUNGSMET published by HERRMANN et al. (1978), but is modified and supplemented (Fig. 31).

The computational scheme was first worked out in BRAITSCH (1962, 1971). It employs the solution equilibria for marine salt systems already described by VAN'T HOFF, D'ANS, and AUTENRIETH and involves solving linear equations of arbitrary degree with the help of determinants. For this work the method of Gauß elimination was applied to coefficient matrices. Since most of the matrices used are weak (i.e., many

| TITLE | : | Project S & L |
|---|---|---|
| Date of sampling | : | 1-24-1980 |
| Region of deposit | : | Niedersachsen |
| Mine/evaporite body | : | Gorleben |
| Plant | : | - |
| Stratigraphical horizon | : | Staßfurt rock salt |
| Locality | : | - |
| Drilling | : | Go 1003 |
| Temperature | : | - |
| pH-value | : | - |
| Flow rate | : | total 11,87 m3 |
| Density of the solution | : | 1.312 g/cm3 at 20 °C |

| Major comp. cations + anions | Mass fraction in % | g/l | mol/ 1000 mol H2O | Major comp. fictitious components | Mass fraction in % | g/l | mol/ 1000 mol H2O | | Ion-% (after Jänecke, only for quinary system) | | Classification of the solution Main group | sub group |
|---|---|---|---|---|---|---|---|---|---|---|---|---|
| Na | 0.3811 | 5.00 | 4.33 | NaCl(2NaCl) | 0.9688 | 12.71 | 4.37 ( | 2.19 ) | 2K | 1.06 | 5 | C |
| K | 0.2683 | 3.52 | 1.79 | KCl (2KCl) | 0.5116 | 6.71 | 1.81 ( | 0.90 ) | Mg | 96.96 | | |
| Mg | 7.6220 | 100.00 | 81.89 | MgCl2 | 29.2482 | 383.74 | 81.00 | | SO4 | 1.98 | | |
| Ca | 0.0534 | 0.70 | 0.35 | Na2SO4 | | | | | Total | 100.00 | | |
| Cl | 21.9436 | 287.90 | 161.63 | K2SO4 | | | | | 2Na | 2.56 | | |
| SO4 | 0.7432 | 9.75 | 2.02 | MgSO4 | 0.7708 | 10.11 | 1.69 | | H2O | 1173 | | |
| CO3 | | | | CaSO4 | 0.1814 | 2.38 | 0.35 | | | | | |
| | | | | CaCl2 | | | | | | | | |
| | | | | CaCO3 | | | | | | | | |
| Total of dissolved salts | 31.0116 | 406.87 | | Total of dissolved salts | 31.6808 | 415.65 | | | | | | |
| Total H2O | 68.9884 | 905.13 | | Total H2O | 68.3192 | 896.35 | | | | | | |
| Total | 100.0000 | 1312.00 | | Total | 100.0000 | 1312.00 | | | | | | |
| | | | | Total Cl in components | 22.6121 | | | | | | | |

| molar fraction in % (mol-%) | | Mol equivalents | | | Equivalent percentages | | Minor components | g/g (ppm) | g/l | Isotope data in per mille |
|---|---|---|---|---|---|---|---|---|---|---|
| Cations: | | Cations | | | Cations | | Li | 50.00 | 0.0656 | D34S |
| | | | | | | | Rb | | | D18O |
| Na + K | 6.93 | Na | 0.0166 | | Na + K | 3.59 | Cs | | | D D |
| Mg | 92.68 | K | 0.0069 | | Mg | 96.00 | NH4 | | | |
| Ca | 0.39 | Mg | 0.6272 | | Ca | 0.41 | Sr | | | |
| | | Ca | 0.0027 | | | | Ba | | | |
| | | | | | | | B | | | |
| Total | 100.00 | Total | 0.6533 | | Total | 100.00 | F | | | |
| | | | | | | | Br | 5350.00 | 7.0192 | |
| | | | | | | | J | | | |
| Anions: | | Anions: | | | Anions: | | P | | | |
| | | | | | | | Mn | | | |
| Cl | 98.77 | Cl | 0.6189 | | Cl | 97.56 | Fe | | | |
| SO4 | 1.23 | SO4 | 0.0155 | | SO4 | 2.44 | Cu | | | |
| CO3 | | CO3 | | | CO3 | | Zn | | | |
| | | | | | | | Sn | | | |
| Total | 100.00 | Total | 0.6344 | | Total | 100.00 | Mg | | | |
| | | | | | | | Pb | | | |
| | | | | | | | Cd | | | |
| | | Difference | 0.0189 | | | | Tl | | | |
| | | Cat.[%] | 2.89 | | | | | | | |
| | | An.[%] | 2.98 | | | | | | | |

**Fig. 30** Printout of the subroutine CHARACTERISTIC VALUES OF SOLUTION ANALYSES.

*Entries*

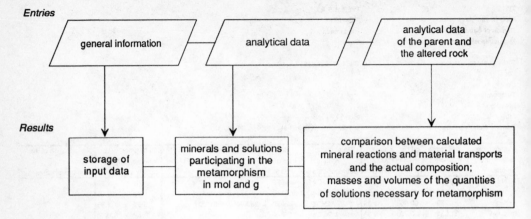

*Results*

**Fig. 31** Structure of the subroutine MINERAL REACTIONS for calculating solution-metamorphism processes.

coefficients equal to 0), a Pivot test has to be conducted. An example of this calculation is given in Fig. 32.

## Computerized evaluation of the calculations

The data processed with the other subroutines are used in this part of the program to calculate the volumes and masses of the altered and/or neogenic evaporites and of the participating solutions. Whereas the other subroutines were able to be automated, this is not feasible in the evaluation phase due to the great number of possible parameters to be considered. Thus, comprehensive knowledge of the internal structure of the program and of the fundamentals of interpreting material balance calculations for mineral reactions in marine evaporite deposits is necessary. The tables of results given in the following were generated with this subroutine.

## 18.2 Salt Rocks

### 18.2.1 Formation of kainitite from carnallitite

In the quinary system of marine evaporites kainite is stable between 15 °C (lower formation temperature) and 83 °C (upper formation temperature). According to BRAITSCH (1962: 93, 139f; 1971: 120f, 181f) kainite on the salt wash surface and salt table must be regarded genetically as a reaction product of solution metamorphism, which is to be considered largely isothermal. An initially kieseritic rock is necessary for kainite to form: the product being a kieseritic carnallitite or a kieseritic Hartsalz (kieserite-sylvite-halite rock). Kainite frequently forms from a carnallitite at and near the salt

| TITLE | : | Project S & L |
|---|---|---|
| Date of sampling | : | . |
| Region of deposit | : | Niedersachsen |
| Mine/evaporite body | : | Gorleben |
| Stratigraphical horizon | : | StaBfurt potash seam (K2) |
| Locality | : | . |
| Drilling | : | . |
| Classification of the solution | : | . |
| Temperature of metamorphism [°C] | : | 25 |
| Kind of ingressing solution | : | saturated NaCl solution |
| Kind of formed solution | : | solution R25°C of the quinary system |

**Fig. 32** Printout of the subroutine MINERAL REACTIONS.

| Calculation scheme | | | | | | | Ingress. sol. | Formed sol. | Calculation test |
|---|---|---|---|---|---|---|---|---|---|
| | c | ks | k | h | - | - | | | |
| 2NaCl | 0 | 0 | 0 | 0.5 | 0 | 0 | 55.5 | 2.37 | 2.37 |
| 2KCl | 0.5 | 0 | 0.5 | 0 | 0 | 0 | 0 | 1.8 | 1.80 |
| MgCl2 | 1 | 0 | 0 | 0 | 0 | 0 | 0 | 80.47 | 80.47 |
| MgSO4 | 0 | 1 | 1 | 0 | 0 | 0 | 0 | 6.26 | 6.26 |
| CaSO4 | 0 | 0 | 0 | 0 | 0 | 0 | 0 | 0 | 0.00 |
| CaCl2 | 0 | 0 | 0 | 0 | 0 | 0 | 0 | 0 | 0.00 |
| H2O | 6 | 1 | 2.75 | 0 | 0 | 0 | 1000 | 1000 | 1000.00 |

| | Calculated mol minerals and solutions | Calculated g minerals and solutions | Left side of the reaction equation | | | | Right side of the reaction equation | | | |
|---|---|---|---|---|---|---|---|---|---|---|
| | | | Composition of ingressing solution [g] | | Dissolved minerals [g] | | Composition of formed solution [g] | | Formed minerals [g] | |
| c | 80.4700 | 22359.07 | | | Totals | | | | | |
| ks | 83.1300 | 11503.36 | NaCl | 4186.82 | c | 22359.07 | NaCl | 277.02 | | 0.00 |
| k | -76.8700 | -18791.59 | KCl | 0.00 | ks | 11503.36 | KCl | 268.40 | | 0.00 |
| h | -66.9041 | -3910.06 | MgCl2 | 0.00 | | 0.00 | MgCl2 | 7661.55 | k | 18791.59 |
| - | 0.0000 | 0.00 | Na2SO4 | 0.00 | | 0.00 | Na2SO4 | 0.00 | h | 3910.06 |
| - | 0.0000 | 0.00 | K2SO4 | 0.00 | | 0.00 | K2SO4 | 0.00 | | 0.00 |
| eindr. | | | MgSO4 | 0.00 | | 0.00 | MgSO4 | 753.45 | | 0.00 |
| Lsg. | 0.6454 | 15813.70 | CaSO4 | 0.00 | | | CaSO4 | 0.00 | | |
| entsl. | | | CaCl2 | 0.00 | | | CaCl2 | 0.00 | | |
| Lsg. | -1.0000 | -26975.42 | H2O | 11626.88 | | | H2O | 18015.00 | | |
| | Normalized to 100 g | Normalized to 100 g on both sides of the reaction equation | Total | 15813.70 | | | Total | 26975.42 | | |
| | c | | Ingr.sol. + dissolved min. | 49676.14 | | | Formed sol. + minerals | 49677.07 | | |
| | | | Normalized values | | | | | | | |
| c | 100.00 | 45.01 | NaCl | 18.73 | c | 100.00 | NaCl | 1.24 | | 0.00 |
| ks | 51.45 | 23.16 | KCl | 0.00 | ks | 51.45 | KCl | 1.20 | | 0.00 |
| k | -84.04 | -37.83 | MgCl2 | 0.00 | | 0.00 | MgCl2 | 34.27 | k | 84.04 |
| h | -17.49 | -7.87 | Na2SO4 | 0.00 | | 0.00 | Na2SO4 | 0.00 | h | 17.49 |
| - | 0.00 | 0.00 | K2SO4 | 0.00 | | 0.00 | K2SO4 | 0.00 | | 0.00 |
| - | 0.00 | 0.00 | MgSO4 | 0.00 | | 0.00 | MgSO4 | 3.37 | | 0.00 |
| Ingress. | | | CaSO4 | 0.00 | | | CaSO4 | 0.00 | | |
| sol. | 70.73 | 31.83 | CaCl2 | 0.00 | | | CaCl2 | 0.00 | | |
| Formed | | | H2O | 52.00 | | | H2O | 80.57 | | |
| sol. | -120.65 | -54.30 | Total | 70.73 | | | Total | 120.65 | | |
| | | | Ingr.sol. + dissolved min. | 222.17 | | | Formed sol. + minerals | 222.18 | | |

Conversion into analyzed composition

| | 1 | 2 | 3 | 4 Conversion of the normalized values into actual composition (column 1) [g] | 5 Calculated composition of the metamorphized salt rock [g] | 6 | 7 | 8 |
|---|---|---|---|---|---|---|---|---|
| Analytical data of the parent rock [mass fraction in %] | | | | | | Metamorphized salt rock [mass fraction in %] | | |
| Actual composition | | Variation | | negative values = right side of the reaction equation | | Calculated composition | (Analyzed) Actual composition | Difference (cc - ac) |
| | | min | max | | | | | |
| c | 25.00 | | | c | 25.00 | c | 0.00 | c | 0.00 | c | 0.00 | 0.0 |
| ks | 16.00 | | | ks | 12.86 | ks | 3.14 | ks | 3.59 | ks | 0.00 | 3.6 |
| k | 0.00 | | | k | -21.01 | k | 21.01 | k | 24.01 | k | 0.00 | 24.0 |
| h | 57.00 | | | h | -4.37 | h | 61.37 | h | 70.12 | h | 0.00 | 70.1 |
| - | | | | - | 0.00 | - | 0.00 | - | 0.00 | - | | 0.0 |
| - | | | | - | 0.00 | - | 0.00 | - | 0.00 | - | | 0.0 |
| a | 1.00 | | | a | | a | 1.00 | a | 1.14 | | | 1.1 |
| - | | | | - | | - | 0.00 | - | 0.00 | - | | 0.0 |
| - | | | | - | | - | 0.00 | - | 0.00 | - | | 0.0 |
| - | | | | - | | - | 0.00 | - | 0.00 | - | | 0.0 |
| ir | 1.00 | | | ir | | ir | 1.00 | ir | 1.14 | ir | 0.00 | 1.1 |
| Total | 100.00 | | | Ingressing solution | 17.68 | Total | 87.52 | Total | 100.00 | Total | 0.00 | |
| Mass fraction [%] of which is dissolved or formed resp. | | c | 25 | formed solution | -30.16 | | | | |
| | | | | Total | 0.00 | | | | |

wash surface of salt domes (e.g., Gorleben). Kainite formation from kieserite + sylvite was often observed on the salt table of the Zechstein deposits in the Werra region (e.g., BRAITSCH 1962: 140, 1971: 182; KÜHN 1957; ROTH 1957). The kainitization of a kieserite-sylvite rock begins along the grain boundaries of the kieserite. In the final stage of formation all kieserite and sylvite is replaced by fine nonoriented fibers of kainite (BRAITSCH 1962: 140, 1971: 182).

In the Werra region kainite also occurs frequently in the vicinity of basalt dikes, which was referred to as »thermal-water kainite« by D'ANS. There is no petrographic difference between this kainite and caprock kainites. However, it must be assumed that, besides their differing ages, kainite also formed near basalt under the influence of a thermal gradient (i.e., polythermal), and not just isothermal conditions (GUTSCHE 1987; GUTSCHE & HERRMANN 1988).

The formation of kainite during solution metamorphism is sometimes character- ized by the occurrence of further secondary minerals: e.g., kainite + bloedite in the form of thick crusts in kieserite-bearing langbeinite in the Ischler evaporite rocks (H. MEYERHOFER 1955 in BRAITSCH 1962: 140, 1971: 183). These mineral reactions can be attributed to the solution metamorphism of a kieseritic langbeinite. In this case the formation temperature of kainite was between 28°C and 37°C (BRAITSCH 1962: 140, 1971: 183).

The caprocks of salt domes display a uniform sequence of various rock types from the outer zones into the salt dome itself (BRAITSCH 1962: 139, 1971: 181). Gypsum- bearing rocks occur outside the dome as reaction products of older anhydrite. They are usually over 100 m thick, highly fissured, and often collapsed, brecciated, filled with solutions, and thus are dreaded during the sinking of shafts. Gypsum caprock fre- quently borders on relatively horizontal boundary surfaces of the evaporite body (hence, salt wash surface). However, the salt wash surface over a salt dome can also have high relief (e.g., Gorleben).

Above potash deposits which intersect the salt wash surface, the kainite caprock usually begins with a thin crust of schoenite, which is sometimes one to several meters thick. Kainite always underlies schoenite, which occasionally forms clusters.

The kainite caprock normally is 20-50 m thick: the kainite is mostly white with the appearance of sugar grains. The rock of the kainite cap consists of a fine-grained mixture of kainite and halite in variable proportions. The original bedding is faint or no longer visible. Only thicker beds of rock salt are preserved. The kainite caprock sometimes contains leonite or bloedite as accessory minerals.

When investigating kainite from zones around the salt wash surface or the salt table, attention is to be paid to secondary minerals such as bloedite, schoenite, leonite, blue rock salt (coarse crystalline?), and sylvite clusters and relict minerals such as kieserite, halite, and sylvite.

## 18.2.2 Formation of K-Mg-free rocks from kainitite

The lack of economic minerals such as carnallite, sylvite, and/or kainite in a potash seam which originally contained these minerals is referred to as an impoverishment. The dissolution, decomposition, and washing away of the K-minerals originally present in the seam frequently occurred under the influence of unsaturated aqueous solutions on the former potash salts. After the aforementioned K-minerals have been dissolved out, the impoverishment consists of rock salt, minute amounts of water-insoluble mineral fractions (quartz, clay minerals) and occasionally anhydrite. These portions of a potash salt horizon are still regarded as impoverished zones even when minute amounts of polyhalite, langbeinite, and/or secondary K-Na, K-Mg, and Na-Mg-minerals are present in the halite rock.

When calculating and quantifying the mineral reactions which led to the imporverishments, a distinction must be made between sulfate-type and chloride-type of marine evaporites.

### Cloride-type of marine evaporites

The simplest form of impoverishment is the alteration of $MgSO_4$-free potash salt of the chloride type into rock salt. The original rock consisted of sylvinite (sylvite-halite rock) which already formed either from carnallitite during solution metamorphism or as primary crystals from $MgSO_4$-depleted seawater. Under the influence of water or NaCl-saturated solutions on the sylvite-halite rocks, the sylvite is completely dissolved and halite crystallizes (due to the NaCl-saturated solutions). NaCl-KCl-saturated solutions form during these processes. No other specific secondary minerals besides halite form at temperatures up to 100 °C and above with chloride-type mineral associations. This is the fundamental difference between the mineral reactions producing impoverishments in chloride-type rocks and the mineral reactions occurring in sulfate-type potash salts.

The anhydrite and the water-insoluble minerals (residues, a + ir) contained in the original rock are enriched relatively and absolutely in the impoverished rocks during alteration (the volume of impoverished rock is less than the original; BRAITSCH 1960, 1962: 95, 1971: 122; HOFFMANN 1961). If the primary potash salt was carnallitite and the NaCl-saturated metamorphic solutions also contained dissolved $CaSO_4$, the $CaSO_4$ will precipitate out. The crystallizing calcium sulfate, as 2nd-generation anhydrite, together with 1st-generation anhydrite already contained in the initial carnallitite, then becomes a part of the impoverishment during the alteration of the sylvinite which formed from the carnallitite (BRAITSCH 1962: 95, 1971: 122). Polyhalite cannot form from anhydrite because the system lacks the necessary $MgSO_4$ components.

Tab. 15 shows examples of reactions for the removal of sylvite and consequently the temperature-dependent conversion of a KCl-bearing rock into rock salt. It is obvious that in the NaCl-KCl-$H_2O$ system at 25 °C more NaCl-saturated solution is necessary than at 83 °C to dissolve the same amount of sylvite.

**Tab. 15** Dissolution of sylvite by NaCl-saturated solution depending on temperature in chloride-type marine evaporites. sy, sylvite; h, halite (from BRAITSCH 1962: 95, 1971: 122).

| Temperature [°C] | Ingressing saturated NaCl solution [g] | Solid phases dissolved [g] | Solid Phases formed [g] | Amount of KCl - NaCl saturated solution formed [g] |
|---|---|---|---|---|
| 83 | 445.4 | 100 sy | 34.1 h | 511.3 |
| 55 | 574.4 | 100 sy | 35.9 h | 638.5 |
| 25 | 834.2 | 100 sy | 37.4 h | 896.8 |

## Sulfate-type of marine evaporites

The mineral reactions taking place during the conversion of $MgSO_4$-bearing potash salts of the sulfate type into rock salt and other secondary minerals of impoverished zones are considerably more complicated and diverse. In contrast to the chloride type, these mineral reactions are very dependent on reaction temperatures up to 100 °C and on the composition of solutions affecting the K-Mg minerals.

Several examples of the possible mineral reactions for the 25°C isotherm will now be considered (Fig. 33). During the metamorphism of a carnallitic rock by an NaCl-saturated solution, a mineral association of either kainite + halite + kieserite or carnallite forms depending on whether there was an excess of kieserite or carnallite in the initial rock. The $R_{25°C}$ solution is formed. Additional NaCl-saturated solution dilutes the $R_{25°C}$ solution. The composition of the solution moves away from point $R_{25°C}$ along the crystallization path in the Y, X, and W direction when an excess of kieserite was present in the initial rock (case I). If, however, carnallite still remains in a kieseritic carnallitite after conversion of all kieserite into kainite, the composition of the solutions moves from point $R_{25°C}$ via $Q_{25°C}$ along the crystallization path toward $P_{25°C}$ due to the addition of NaCl-saturated solutions (case II).

Tab. 16 shows the composition of the constant solutions in the quinary system. In Tab. 17 the possible mineral reactions for cases I and II of the 25°C isotherms are given together with the calculated material exchange.

The described processes occur in the upper (e.g., salt dome) or lateral (e.g., flat layered evaporite beds) margins of evaporite bodies during, for example, the alteration of a kieseritic carnallitite. The carnallitite is first converted into a kainitic rock by NaCl-saturated solutions. Subsequently, kainite and sylvite dissolve due to the addition of more NaCl solution and mixing with the respectively more concentrated solutions. This produces further K-Mg, Na-K, and Na-Mg sulfates.

After the reactions have ceased, kainite overlies followed by an impoverishment (Fig. 23) the unaltered carnallitite.

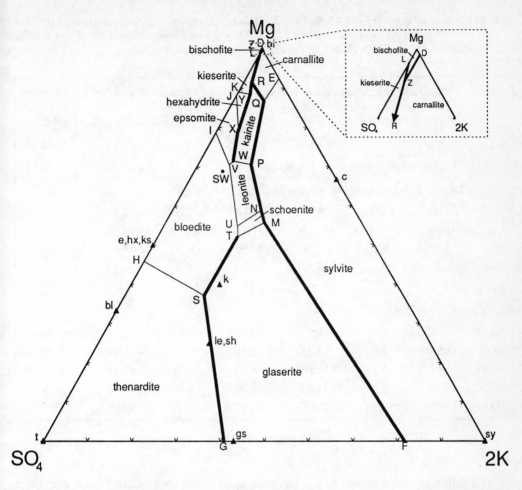

**Fig. 33** The quinary system of marine evaporites with NaCl saturation and stable equilibria (25°C). The thick lines represent crystallization paths, the thin transition lines. The constant solutions are a comparison of values from D'Ans (1933), Braitsch (1962, 1971), Harvie & Weare (1980), Gudowius (1984), and Herrmann (unpublished compilation). See text for further explanation.

## 18.2.3 Quantification of the reactions carnallitite → kainitite → K-Mg-free rock

Since all previously available data on the geology and chemical composition (rocks, solutions) of the Gorleben salt dome were obtained from exploratory drilling and the study of drill cores, more detailed data can only be obtained with continued subsurface exploration. Hence, the computational results presented in the following are to be considered a first attempt at quantifying the material reactions and transport which occurred in the geological past.

**Tab. 16** Constant solutions in the NaCl-KCl-MgCl$_2$-MgSO$_4$-H$_2$O system with NaCl saturation (univariant points) and NaCl-saturated solution (25 °C). Unpublished compilation after D'Ans (1933), Braitsch (1962, 1971), Harvie & Weare (1980), and Gudowius (1984).

| Point | Saturation of the solution with NaCl, NaMg- and Mg-sulfates Crystallization path | | | | Saturation with NaCl and KCl Crystallization path | | Saturated NaCl solution |
|---|---|---|---|---|---|---|---|
| | R | Y | X | W | Q | P | - |
| Author | Gu | Br | H&W | Gu | Gu | Gu | DA |
| Data in Jänecke formula units [2K + Mg + SO$_4$ = 100] | | | | | | | |
| 2K | 1.90 | 3.00 | 5.00 | 7.60 | 7.00 | 12.10 | - |
| Mg | 91.50 | 88.50 | 78.80 | 71.50 | 86.80 | 70.50 | - |
| SO$_4$ | 6.60 | 8.50 | 16.20 | 20.90 | 6.20 | 17.40 | - |
| 2Na | 2.50 | 3.10 | 7.60 | 11.80 | 5.50 | 14.50 | - |
| H$_2$O | 1 055 | 1 056 | 1 100 | 1 124 | 1 186 | 1 236 | - |
| Data in mol/1000 mol H$_2$O | | | | | | | |
| 2NaCl | 2.37 | 2.93 | 6.91 | 10.50 | 4.89 | 11.73 | 55.50 |
| 2KCl | 1.80 | 2.84 | 4.55 | 6.76 | 6.22 | 9.79 | 0.00 |
| MgCl$_2$ | 80.47 | 75.76 | 56.91 | 45.02 | 71.71 | 42.96 | 0.00 |
| MgSO$_4$ | 6.26 | 8.05 | 14.73 | 18.59 | 5.52 | 14.08 | 0.00 |
| H$_2$O | 1 000 | 1 000 | 1 000 | 1 000 | 1 000 | 1 000 | 1 000 |

The following data served as the basis for the calculations of material transport in the area of the salt wash surface of the Gorleben salt dome (Tätigskeitsbericht der BGR 1985/86; Bornemann et al. 1988; see also Chapter 17):
- Staßfurt potash seam with a thickness of 20-30 m, averaging 25 m,
- initial composition of the brecciated carnallitite:

| | |
|---|---|
| halite | 57 % |
| carnallite | 25 % |
| kieserite | 16 % |
| anhydrite + water-insoluble residues | 1 % |

This initial rock was converted into halite (impoverishment) down to an average of 110 m below the salt wash surface. The influence of unsaturated aqueous solutions can be detected down to about 155 m below the salt wash surface. Consequently, a kainitization of the Staßfurt seam was assumed for the depth range of 110-155 m (difference = 45 m) below the salt wash surface for the model calculations (cf. Fig. 23).

**Tab.17** Possible mineral reactions for cases I and II at 25°C (see text for explanation). The values given represent material conversions in g, normalized to 100 g of the respective mineral. The reaction numbers given in parentheses represent intermediate steps which were calculated without matrices. c, carnallite; ks, kieserite; k, kainite; h, halite; hx, hexahydrite; e, epsomite; sy, sylvite.

| Case | Reaction no. | Left side of reaction equation | | | Right side of reaction equation | |
|---|---|---|---|---|---|---|
| | | Ingressing solution [g] | Existing solution [g] | Existing solid phases [g] | Solid phases formed [g] | Solution formed [g] |
| Case I | 1 | NaCl-Lsg. | - | c + ks | k + h | $R_{25°C}$ |
| | (= case II) | 70.73 | - | 100 + 51.45 | 84.04 + 17.49 | 120.65 |
| | 2a | NaCl-Lsg. | $R_{25°C}$ | k | ks + h | $Y_{25°C}$ |
| | | 229.13 | 4535.21 | 100 | 3.32 + 46.08 | 4814.94 |
| | (2b) | NaCl-Lsg. | - | ks | hx + h | - |
| | | 88.53 | - | 100 | 165.09 + 23.44 | - |
| | 2(a + b) | NaCl-Lsg. | $R_{25°C}$ | k | hx + h | $Y_{25°C}$ |
| | | 232.07 | 4535.21 | 100 | 5.48 + 46.86 | 4814.94 |
| | 3 | NaCl-Lsg. | $Y_{25°C}$ | k + hx | h | $X_{25°C}$ |
| | | 440.51 | 1714.23 | 100 + 74.44 | 70.04 | 2259.13 |
| | (4) | NaCl-Lsg. | - | hx | e + h | - |
| | | 10.73 | - | 100 | 107.89 + 2.84 | - |
| | 5 | NaCl-Lsg. | $X_{25°C}$ | k + e | h | $W_{25°C}$ |
| | | 296.87 | 1366.46 | 100 + 9.83 | 40.52 | 1732.64 |
| Case II | 1 | NaCl-Lsg. | - | c + ks | k + h | $R_{25°C}$ |
| | (= case I) | 70.73 | - | 100 + 51.45 | 84.04 + 17.49 | 120.65 |
| | 2 | NaCl-Lsg. | - | c + k | sy + h | $Q_{25°C}$ |
| | | 68.20 | - | 100 + 6.77 | 24.24 + 15.19 | 135.54 |
| | 3 | NaCl-Lsg. | $Q_{25°C}$ | k + sy | h | $P_{25°C}$ |
| | | 345.44 | 614.32 | 100 + 3.83 | 52.40 | 1011.20 |

In the reaction sequence, anhydrite (a) and water-insoluble minerals (insoluble residues = ir) were neglected. The values for mineral densities necessary for the calculation of volume were taken from WOHLENBERG (1982).

The calculated material conversions are given in Tabs. 18 and 19 and Figs. 34 and 35. Since no chemical analyses were available for the rocks which formed (impoverished potash seam forming rock salt, kainitite), the impoverishment of the carnallitic rock was calculated assuming complete removal of K-Mg-$SO_4$. For the same reason it was presumed for kainitization that the carnallite of the initial rock was fully decomposed. The figures show that for kainitization of the carnallitic rock substantially less ingressing NaCl-saturated solution is necessary than for the formation of completely K-Mg-free rocks from carnallitite.

Before balancing the impoverishment reactions (Tab. 18) the kainitization had to be calculated, as given in Tab. 19 for depths below the zone of impoverishment. As follows from Tab. 17 (case I, reaction 2a and 2a+b), in the next reaction step the forming kainitite is affected by the $R_{25°C}$ solution, which formed during the conversion of the carnallitite and additional NaCl-saturated solution (Fig. 33; Tab. 17, case I, reaction 2a and 2a + b). However, the recalculation of the rock composition shows that only about 3% of the mass of the $R_{25°C}$ solution necessary for the formation of the $Y_{25°C}$ solution in the second reaction step is formed in the first reaction (kainitization). Consequently, about 97% of the kainite forming in the first calculation step is still present in addition to kieserite. Since, however, there are no data on the solubility of $MgSO_4$ in the $NaCl$-$KCl$-$MgSO_4$-$H_2O$ system, a kainite-saturated solution was assumed for subsequent calculations (i.e., dissolution of the remaining kieserite; H.-E. USDOWSKI, personal communication). This is a simplification due to the appearance of reciprocal salt pairs in the $KCl$-$MgSO_4$-$H_2O$ and $NaCl$-$MgSO_4$-$H_2O$ systems. Neglecting the secondary constituents (a + ir) is another uncertainty in the calculations. Regarding the order of magnitude, however, the computational results are comparable with the material transports occurring in nature.

Deviations from previous computational results could arise if more accurate data for both the initial rock (brecciated carnallitite) and the rock formed (kainitite, rock salt) were available. For example, impoverishment was calculated for complete removal of the K-Mg-$SO_4$ contents of the initial rock. Thus, in these calculations only anhydrite, water-insoluble-residue, and above all halite still remain in the newly formed rock (Tab. 18). However, in the Gorleben salt dome the new rock may locally occur in the form of kieseritic rock salt. The ratio of ingressing solution to initial rock - and thus of forming solution to forming rock - shown in Fig. 34 would decrease assuming an incomplete dissolution of the kieserite.

Tab. 18 Balance calculation for the formation of K-Mg-free rocks (impoverishment) from carnallitic rock following the conversion of a carnallitite into a kainitic rock at 25°C by a NaCl-saturated solution. Examples 1-3 show the results for the differing horizontal extension of the Staßfurt potash seam at depths of 0-110 m below the salt wash surface. The average thickness of the original carnallitite was 25 m. In example 4, the thickness of the original carnallitite was assumed to be 25 m based on the thickness of the halite rock (rock salt) which formed. c, carnallite; ks, kieserite; h, halite; a, anhydrite; ir, insoluble residues.

Initial rock: carnallitite

| Example | Seam dimensions [m] | | | Rock | | Composition Mass [t] | | | | Ingressing solution: NaCl-saturated solution | |
|---|---|---|---|---|---|---|---|---|---|---|---|
| | Thickness | Depth | Horizontal extension | Volume [m³] | Mass [t] | c | ks | h | a + ir | Volume [m³] | Mass [t] |
| 1 | 25 | 45 | 10 | 11 250 | 23 643 | 5 911 | 3 783 | 13 477 | 473 | 30 110 | 36 132 |
| 2 | 25 | 45 | 100 | 112 500 | 236 430 | 59 108 | 37 829 | 134 765 | 4 729 | 301 097 | 361 317 |
| 3 | 25 | 45 | 1 000 | 1 125 000 | 2 364 300 | 591 075 | 378 288 | 1 347 651 | 47 286 | 3 010 975 | 3 613 170 |
| 4 | 37.2 | 45 | 1 000 | 1 674 000 | 3 518 078 | 879 520 | 562 893 | 2 005 305 | 70 362 | 4 480 330 | 5 376 397 |

Rock formed: rock salt

| Example | Estimated thickness of seam at same depth and horizontal extension as initial rock | Rock | | Composition Mass [t] | | Total solution formed | |
|---|---|---|---|---|---|---|---|
| | | Volume [m³] | Mass [t] | h | a + ir | Volume [m³] | Mass [t] |
| 1 | 16.8 | 7 575 | 16 536 | 16 061 | 475 | 34 460 | 43 243 |
| 2 | 16.8 | 75 748 | 165 357 | 160 605 | 4 751 | 344 605 | 432 433 |
| 3 | 16.8 | 757 479 | 1 653 567 | 1 606 055 | 47 512 | 3 446 045 | 4 324 329 |
| 4 | 25.0 | 1 127 129 | 2 460 507 | 2 389 809 | 70 698 | 5 127 715 | 6 434 602 |

**Tab. 19** Balance calculation for the conversion of a carnallitite into a kainitic rock at 25°C due to a NaCl-saturated solution. Examples 1-3 show the results for the differing horizontal extension of the Staßfurt potash seam at depths of 110-155 m below the salt wash surface. The average thickness of the original carnallitite was 25 m. In example 4, the thickness of the original carnallitite was assumed to be 25 m based on the thickness of the kainitic rock which formed. c, carnallite; ks, kieserite; h, halite; a, anhydrite; ir, insoluble residues; k, kainite.

Initial rock: carnallitite

| Example | Seam dimensions [m] | | | Rock | | Composition Mass [t] | | | | Ingressing solution: NaCl-saturated solution | |
|---|---|---|---|---|---|---|---|---|---|---|---|
| | Thickness | Depth | Horizontal extension | Volume [m³] | Mass [t] | c | ks | h | a+ir | Volume [m³] | Mass [t] |
| 1 | 25 | 45 | 10 | 11 250 | 23 643 | 5 911 | 3 783 | 13 477 | 473 | 3 484 | 4 181 |
| 2 | 25 | 45 | 100 | 112 500 | 236 430 | 59 108 | 37 829 | 134 765 | 4 729 | 34 839 | 41 807 |
| 3 | 25 | 45 | 1 000 | 1 125 000 | 2 364 300 | 591 075 | 378 288 | 1 347 651 | 47 286 | 348 389 | 418 067 |
| 4 | 29,7 | 45 | 1000 | 1 336 950 | 2 809 734 | 702 434 | 449 557 | 1 601 548 | 56 195 | 414 026 | 496 831 |

Rock formed: kainitite

| Example | Estimated thickness of seam at same depth and horizontal extension as initial rock | Rock | | Composition Mass [t] | | | | Total solution R$_{25°C}$ formed | |
|---|---|---|---|---|---|---|---|---|---|
| | | Volume [m³] | Mass [t] | k | ks | h | a+ir | Volume [m³] | Mass [t] |
| 1 | 21,03 | 9 466 | 20 692 | 4 967 | 742 | 14 510 | 473 | 5 444 | 7 131 |
| 2 | 21,03 | 94 656 | 206 924 | 49 674 | 7 424 | 145 097 | 4 729 | 54 438 | 71 313 |
| 3 | 21,03 | 946 563 | 2 069 235 | 496 739 | 74 239 | 1 450 971 | 47 286 | 544 376 | 713 132 |
| 4 | 25,00 | 1 124 895 | 2 459 079 | 590 325 | 88 226 | 1 724 334 | 56 195 | 646 936 | 847 486 |

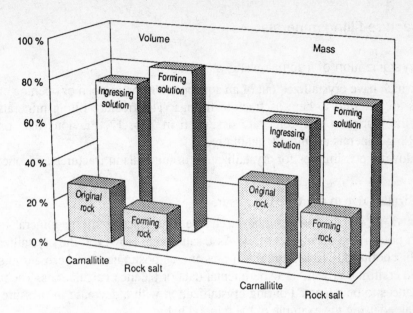

**Fig. 34** Results for the calculation of the formation model for a K-Mg-free rock (impoverishment) from carnallitite in the Staßfurt potash seam of the Gorleben salt dome at 25°C. The totals of original rock + ingressing solution and forming rock + forming solution were recalculated to 100 %.

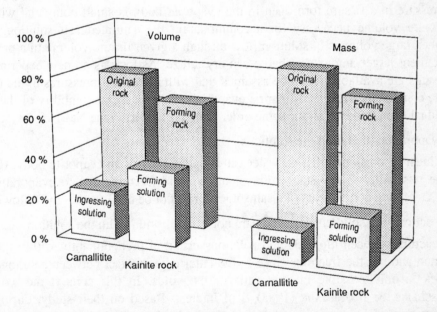

**Fig. 35** Results for the calculations of the model of the conversion of a carnallitite in a kainitic rock in the Staßfurt potash seam of the Gorleben salt dome at 25°C. The totals of original rock + ingressing solution and forming rock + forming solution were recalculated to 100 %.

## 18.3　Fracture-filling minerals

### 18.3.1 Crystallization of fracture-filling minerals

Minerals which have crystallized out of an aqueous solution in open or opening cracks or fissures are referred to here as fracture fillings. The fracture-filling minerals observed in the Gorleben salt dome are described in Tab. 12. Previous data on the fracture-filling minerals are mostly qualitative.

The following possibilities for crystallization of minerals in fractures are conceivable:

1. Crystallization due to decrease in pressure

   RICHTER & KLARR (1984) describe bischofite as a fracture-filling mineral in the Staßfurt potash salt seam (K2) of the Asse salt dome. However, the crystallization of the bischofite due to a decrease in pressure as postulated by these authors has not been confirmed by either experimental data or balance calculations. The material balances to be expected during crystallization with a decrease in pressure will be explained using the example of the mineral halite.

   The solubility of NaCl only decreases by about 8 g NaCl/1000 g solution in the $NaCl-H_2O$ system at 25°C between 0 and 150 MPa (600 - 700 m depth). At 25°C and with a decrease in pressure from 25 to 0.1 MPa approx. 1.5 g of NaCl crystallizes from 1000 g of solution (ADAMS 1931: 3806; KAUFMANN 1968: 617, Tab. 56). Therefore, the amount of halite which crystallized due to a decrease in pressure in a fissure torn open in the evaporite body is small compared with the fissure volume necessary for accommodating the saturated salt solution. Only when large volumes of solution flow through a given fissure volume in a pressure gradient larger amounts of halite can crystallize. As long as no new experimental results are available, it can be assumed that with increasing pressures in the multiple component systems of marine evaporites the increase in solubility of the individual compounds is on the same order of magnitude as in the $NaCl-H_2O$ system.

2. Evaporation of $H_2O$ of the solutions

   It must be presumed that no water can evaporate out of an evaporite body. Hence, the possibility for crystallization of minerals in fractures due to evaporation of $H_2O$-bearing constituents of a salt solution will not be discussed in the following.

3. Reaction with solid phases which are not in equilibrium with the solution

   Reaction between minerals and solutions occurred predominantly within the salt strata and not the fractures since in the latter case, contact between solutions and rock is only possible on the walls of the voids. In this context the work of FISCHBECK & BORNEMANN (1988) is of interest. Based on their study, during the formation of fracture-filling halite in the Na2 of the Gorleben salt dome the Br content of the adjacent rock salt has not been influenced by the solutions.

This observation agrees with the fact that the rock salt was obviously only slightly affected by the fluid components in the vicinity of basalt dikes and the metamorphic solutions along the boundary between the Staßfurt potash seam and the adjacent Na3 rock salt in Gorleben.

4. Crystallization due to a decrease in temperature

Frequent decreases in temperature were certainly involved in the crystallization of fracture-filling minerals. This is especially true for solutions which were able to move upward along vertical fractures into cooler rock.

The cooling of saturated solutions in the $KCl-H_2O$ and $NaCl-KCl-MgCl_2-H_2O$ systems clearly benefits - in terms of quantity - the formation of sylvite due to the positive temperature coefficients for KCl compared with the crystallization of NaCl (cf. Fig. 23, right).

5. Mixtures of solutions of various origin and composition

A maximum amount of halite crystallizes when a NaCl-saturated solution mixes with a $NaCl-MgCl_2$-saturated solution. This model has practical significance in salt mining when greater quantities of NaCl solutions from the surrounding rock flow into the mine through open paths. Such fluid ingression can theoretically be stopped by crystallization of NaCl when the cracks and fissures acting as paths are healed. For this to happen, the ingressing NaCl solution would have to mix with a $NaCl-MgCl_2$-saturated solution.

Under natural conditions it is conceivable that NaCl-satuarated solutions - from the hanging-wall rocks of the salt dome, for example - mix with $MgCl_2$-bearing solutions to form a carnallitic rock (metamorphic solutions). This can occur in the evaporite body under isothermal or polythermal conditions.

## 18.3.2 Quantification of halite crystallisation in cracks and fissures

In the Gorleben salt dome halite has been observed to be the dominant fracture-filling mineral. Computational models of the crystallization of fracture-filling halite with a decrease in temperature and with mixing of different solutions were developed for interpreting the composition of fracture-filling minerals. The computational models were supplemented and modified as exploration of the salt dome and related mineralogical and geochemical studies progressed.

The computational results for halite formation due to cooling of a solution in the $NaCl-H_2O$ system is shown in Tab. 20 and Fig. 36. A temperature interval of 60°C for the calculated cooling of a NaCl-saturated solution from 85°C - 25°C (Tab. 20) appears to be relatively large for natural conditions in an evaporite body with occasional fissures. As expected, however, the amount of solution necessary for forming a certain amount of halite increases as the selected temperature interval decreases.

The absolute bromide content of the NaCl of the fracture fillings and the adjacent rock (rock salt) is an important criterion for evaluating the genesis of the fracture-filling halite. According to studies of FISCHBECK & BORNEMANN (1988) the bromide contents of the fracture-filling halite in the hanging-wall rock salt Na2γ of the Staßfurt sequence (drilling Go 1005, 126 m below the salt wash surface) varies between 137 and 173 µg Br/g NaCl with an average of 160 µg Br/g NaCl. In the adjacent rock salt the Br contents range 90-130 µg Br/g NaCl. These values clearly evidence that the Br contents in the fracture-filling halite are not able to be explained solely by the dissolution of rock salt (NaCl-$H_2O$ system) from the Staßfurt and/or Leine sequence and renewed crystallization of halite from these solutions (e.g., due to a decrease in temperature). The following calculation is an example:

The unfavorable case is assumed in which halite with a maximum of 300 µg Br/g NaCl is dissolved only from the upper part of the Staßfurt rock salt by Br-free $H_2O$ (NaCl-$H_2O$ system). At 25 °C a NaCl-saturated solution (36 g NaCl dissolved in 100 g $H_2O$) would then have to contain around 80 µg Br/g NaCl solution. The distribution coefficient

$$b = \frac{\% Br_{Mineral}}{\% Br_{L\ddot{o}sung}}$$

is 0.14 ± 0.02 at 25 °C for initial NaCl crystallization from seawater and 0.073 for initial carnallite crystallization (BRAITSCH & HERRMANN 1962; BRAITSCH & HERRMANN 1963; HERRMANN 1972; HERRMANN et al. 1973; HERRMANN 1980). Accordingly, the halite crystallizing from a NaCl solution with 80 µg Br/g NaCl must contain 80 · 0.14 = 11 µg Br/g NaCl or 80 · 0.073 = 6 µg Br/g NaCl. These expected bromide contents are one order of magnitude less than those actually observed. This means that the fracture-filling halite did not form from the dissolution of Staßfurt or Leine rock salt in Br-free water with subsequent NaCl crystallization due to a decrease in temperature.

The Br contents of 137-173 µg Br/g halite observed in fracture-filling halite must have crystallized from solutions with 1100-2200 µg Br/g solution (the varying Br contents correspond to the varying distribution coefficients). Salt solutions with such high Br contents correspond with either concentrated seawater (e.g., BRAITSCH 1962: 106f, 1971:138) or metamorphic solutions with high $MgCl_2$ contents from the decomposition of primary carnallitite with Br contents of about 3000 µg Br/g carnallite. There have as yet been no reliable studies of the Br content in the carnallite of the carnallitite from the Staßfurt potash seam of the Gorleben salt dome. During the decomposition of a carnallitite of the average composition of the Staßfurt potash seam in Gorleben (see Chapters 17.1 and 18.2.3) and an assumed Br content of 3000 µg Br/g carnallite, a $R_{25°C}$ solution forms (cf. Chapters 18.2.1 to 18.2.3) with around 2130 µg Br/g solution. Solutions with equally high and even higher Br contents were in fact discovered in exploratory drillings Go 1002, 1003, and 1005 (see Tab. 14).

The $R_{25°C}$ solution which formed as a result of carnallite decomposition contains $SO_4$ constituents. Since these $SO_4$ constituents can be neglected in the first approximation for the crystallization of fracture-filling halite, mixtures of NaCl-saturated and $MgCl_2$-bearing solutions were used in the model calculations. The latter involves

**Tab. 20** Balance calculations for the crystallization of fracture-filling halite due to cooling of a NaCl-saturated solution.

| Dimensions of fissures [m] | | | Halite | | NaCl solution necessary for cooling from | | | | | |
| Width | Height | Horizontal extension | Volume [m³] | Mass [t] | 85 to 55°C Volume [m³] | 85 to 55°C Mass [t] | 55 to 25°C Volume [m³] | 55 to 25°C Mass [t] | 85 to 25°C Volume [m³] | 85 to 25°C Mass [t] |
|---|---|---|---|---|---|---|---|---|---|---|
| 0.01 | 10 | 10 | 1 | 2.16 | 170 | 200 | 319 | 379 | 112 | 132 |
| 0.01 | 50 | 50 | 25 | 54.00 | 4 259 | 5 004 | 7 972 | 9 487 | 2 798 | 3 288 |
| 0.10 | 10 | 10 | 10 | 21.60 | 1 703 | 2 002 | 3 189 | 3 795 | 1 119 | 1 315 |
| 0.10 | 50 | 50 | 250 | 540.00 | 42 586 | 50 038 | 79 724 | 94 871 | 27 985 | 32 882 |

| Residual NaCl solution for cooling from | | | | | |
| 85 to 55°C Volume [m³] | 85 to 55°C Mass [t] | 55 to 25°C Volume [m³] | 55 to 25°C Mass [t] | 85 to 25°C Volume [m³] | 85 to 25°C Mass [t] |
|---|---|---|---|---|---|
| 166 | 198 | 314 | 377 | 108 | 129 |
| 4 160 | 4 950 | 7 861 | 9 433 | 2 695 | 3 234 |
| 1 664 | 1 980 | 3 144 | 3 773 | 1 078 | 1 294 |
| 41 595 | 49 498 | 78 609 | 94 331 | 26 952 | 32 342 |

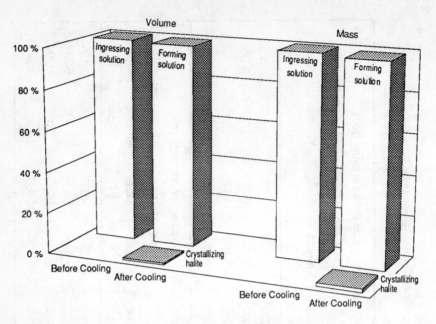

**Fig. 36** Results for the calculation of the crystallization of fracture-filling halite due to cooling of a NaCl-saturated solution from 85°C to 25°C in the Gorleben salt dome. The volumes and masses of ingressing solution and crystallizing halite + forming solution were recalculated to 100%.

simpler compositions of point E (saturation with carnallite) and D (saturation with bischofite) of the quaternary $NaCl-KCl-MgCl_2-H_2O$ system. The Br contents discussed here also apply to solutions of the quinary system.

Halite crystallizing from solutions with very high $MgCl_2$ contents and about 2130 µg Br/g solution at 25°C contain around 155 µg Br/g NaCl ($b_{halite} = 0.073$). A $MgCl_2$-bearing solution with 1620 µg Br/g solution forms when three parts of a $MgCl_2$ solution with 2130 µg Br/g solution mix with one part of a NaCl-saturated solution with an assumed 80 µg Br/g solution. This mixing model involving $MgCl_2$ solution ($E_{25°C}$) + NaCl solution will be discussed below as a possibility for the formation of fracture-filling halite. The model for the $D_{25°C}$ solution (with a $MgCl_2$ solution/NaCl solution ratio of 1/1) was also calculated for comparison.

The halite crystallizing from the mixing of NaCl and $MgCl_2$ solutions must then contain $1620 \cdot 0.073 = 120$ µg Br/g halite. This expected Br content corresponds to the lower boundary for the values observed in fracture-filling halite. It increases when calculated for the solutions with higher Br contents from deep drillings (Tab. 14). Consequently, the Br contents of the fracture-filling halite support the mixing model of $MgCl_2$ solution + NaCl-saturated solution. This possibility for the formation of fracture-filling halite in the Gorleben salt dome has also been discussed in FISCHBECK & BORNEMANN (1988).

The balances for the isothermal mixing models (at 25°C) are given in Tab. 21. Tab. 21 and Figs. 37 and 38 show clearly that less halite crystallizes due to mixing of a NaCl-saturated solution and an $E_{25°C}$ solution than due to mixing with a $D_{25°C}$ solution. The latter is in equilibrium with the minerals carnallite and bischofite. However, bischofite, which occurs relatively seldom in Zechstein salt deposits has not yet been observed in the Gorleben salt dome. However, after a new model primary bischofite has been dissolved by seawater and halite crystallized (HERRMANN & v.BORSTEL 1991).

Fig. 39 gives an idea of the solution volume necessary for the crystallization of 1 m³ of fracture-filling halite due to cooling of a NaCl solution or mixing of NaCl and MgCl₂ solutions.

**Tab. 21** Balance calculations for the crystallization of fracture-filling halite due to mixing of a NaCl-saturated solution with a solution D and E at 25°C.

| Dimensions of fissures [m] | | | Crystallized NaCl | | Solutions [t] | | | Solutions [m³] | | |
|---|---|---|---|---|---|---|---|---|---|---|
| Width | Height | Horizontal extension | [m³] | [t] | Saturated NaCl | $D_{25°C}$ | Formed | Saturated NaCl | $D_{25°C}$ | Formed |
| **$D_{25°C}$** | | | | | | | | | | |
| 0.01 | 10 | 10 | 1 | 2.16 | 14.6 | 17.4 | 29.9 | 12.2 | 13.0 | 23.4 |
| 0.01 | 50 | 50 | 25 | 54.000 | 364 | 436 | 747 | 304 | 323 | 586 |
| 0.10 | 10 | 10 | 10 | 21.600 | 146 | 174 | 299 | 121 | 129 | 234 |
| 0.10 | 50 | 50 | 250 | 540.000 | 3 645 | 4 361 | 7 465 | 3 037 | 3 235 | 5 860 |

| Dimensions of fissures [m] | | | Crystallized NaCl | | Solutions [t] | | | Solutions [m³] | | |
|---|---|---|---|---|---|---|---|---|---|---|
| Width | Height | Horizontal extension | [m³] | [t] | Saturated NaCl | $E_{25°C}$ | Formed | Saturated NaCl | $E_{25°C}$ | Formed |
| **$E_{25°C}$** | | | | | | | | | | |
| 0.01 | 10 | 10 | 1 | 2.16 | 19.5 | 63.3 | 80.7 | 16.3 | 49.6 | 65.1 |
| 0.01 | 50 | 50 | 25 | 54.000 | 488 | 1 583 | 2 018 | 407 | 1 239 | 1 628 |
| 0.10 | 10 | 10 | 10 | 21.600 | 195 | 633 | 807 | 163 | 496 | 651 |
| 0.10 | 50 | 50 | 250 | 540.000 | 4 884 | 15 834 | 20 176 | 4 070 | 12 390 | 16 284 |

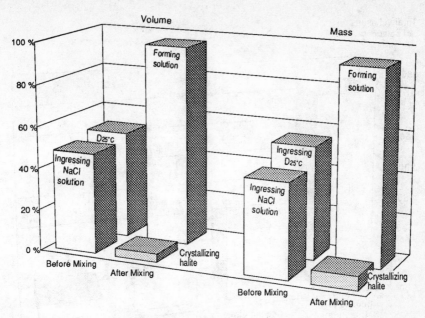

**Fig. 37** Results for the calculation of crystallization of fracture-filling halite due to mixing of a NaCl-saturated solution with a solution D at 25°C. The volumes and masses of mixing solutions and crystallizing halite + forming solution were recalculated to 100%.

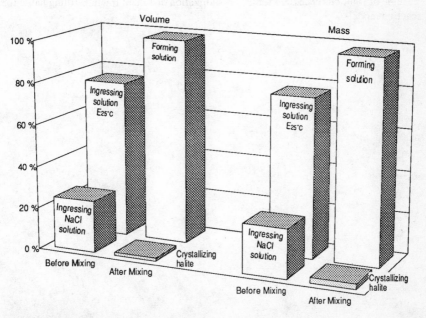

**Fig. 38** Results for the calculation of crystallization of fracture-filling halite due to mixing of a NaCl-saturated solution with an solution E at 25°C. The volumes and masses of mixing solutions and crystallizing halite + forming solution were recalculated to 100%.

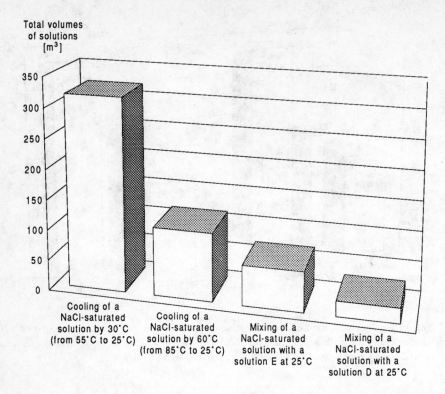

**Fig. 39** Volume of solutions necessary for the crystallization of 1 m³ of fracture-filling halite for various reaction models.

# 19 Evaluation of the current situation

The long-term safety of repositories for radioactive wastes in evaporites can only be adversely affected by the following scenario:

Aqueous solutions react with the wastes, contaminants are dissolved by the fluid phases, the contaminated solutions migrate from the repository through the strata of the salt dome and surrounding and overlying rock up to the biosphere. Hence, in a geological repository system the following observations are of importance to the evaluation of long-term safety:

1. evidence of mineral reactions and material transport between the time of formation of the geological system and today provided by the chemical and mineralogical composition of the rocks and
2. evidence of preferred pathways in the rock.

In the case of Gorleben the current status of exploration only allows methods to be developed and initial studies conducted with observations from aboveground activities. The previous calculations of material conversions and transports are based above all on observations in the upper parts of the salt dome. It remains to be seen whether similar observations (fracture fillings in rock salt beds, formation of the potash seam as carnallitite, Hartsalz, or K-Mg-free salts) are also made for rocks at depths of 800-1000 m depth. The lack of fracture-fillings in the rock salt and the formation of the Staßfurt potash seam as carnallitite combined with other necessary criteria (e.g., Br contents in carnallite and halite) could be evidence that the rocks in deeper parts of the salt dome was indeed deformed during salt dome formation, and that in contrast, the original composition of the rocks were hardly, if at all, altered by solutions. This statement does not apply to the differentiation of various evaporite minerals in the dominantly solid state during deformation (so-called dynamic metamorphism), which is practically insignificant to questions regarding the long-term safety of underground repositories.

In this context, the necessity of continuing the work on gases and solutions in evaporites and on the composition and genesis of secondary minerals in evaporite beds in the course of the underground exploration of the Gorleben salt dome is pointed out again.

The evaluation of possible material transport in the geological past by solutions at greater depth in the Gorleben salt dome will first be able to be determined and quantified during further mining exploration. The appropriate information must first be available before it can be determined whether and to what extent aqueous solutions were effective in the geological past at greater depths in the Gorleben salt dome (today about 800-1000 m deep) and how the actual salt dome is to be evaluated as a barrier for a hazardous waste repository.

All pertinent observations are to be analyzed and quantified as to their extent and effectiveness over the geological past. Finally, the results are to be interpreted regarding their possible consequences for the long-term safety of the repository situated in a geological system. The prerequisite for this is, in addition to site-related criteria, the consideration of non-site-related observations and data for comparison from compositionally similar geological systems (e.g., other salt domes of the same mining district, stratigraphically and petrographically similar evaporites in vertical and flat layered strata).

With these data it will be possible to promote studies like PAGIS (Performance Assessment of Geological Isolation Systems for Radioactive Waste; e.g., CADELLI et al. 1988; CADELLI & COTTONE 1990) and other scenario investigations. To validate these calculation models the quantification of mineral reactions and material transports is necessary. A validation can increase the public acceptance of studies on long-term safety and the final disposal of wastes in general. This acceptance is the only way to discuss new disposal sites, as political and social developments e.g., in the crowded European countries show during the last years.

Under no circumstances, however, are underground repositories only to be regarded for disposal of relatively small quantities of radioactive waste, as is still done today. A similar, but even greater problem involves the long-term safe isolation of much greater volumes of toxic nonradioactive wastes. This problem is much more acute for scientists involved in planning and politicians involved in responsible exercise of power, not only now but far into the future as well.

Scientists, politicians, and citizens interested in constructive solutions must critically control every individual step taken toward the necessary realization of underground repositories. We ourselves will no longer be able to evaluate the effectiveness of the underground repositories we build today: this will be left up to future generations. In other words, it will be the people living in centuries to come who will be confronted with the errors made in underground repository construction today. We must never neglect this aspect during the planning and initial operation of underground repositories.

# 20 References

ACKERMANN, G., SCHRADER, R., HOFFMANN, K. (1964): Untersuchungen an gashaltigen Mineralsalzen, II. Teil: Methodik und Ergebnisse der gasanalytischen Untersuchungen. - Bergakademie, 16: 676-679. Berlin.

ADAMS, L.H. (1931): Equilibrium in binary systems under pressure. I. An experimental and thermodynamic investigation of the system $NaCl-H_2O$ at 25 °C. - J. Am. Chem. Soc. 53: 3769-3813.

AKSTINAT, M. (1983): Untersuchungen an Flüssigkeits- und Gasproben aus der Bohrung Go 5001 und Go 5002. - pp. 9, Bericht des Instituts für Tiefbohrtechnik, Erdöl- und Erdgasgewinnung (ITE), TU Clausthal, für die DBE in 3150 Peine, unpub.

ALBRECHT, H. (1932): Das Erdölvorkommen von Volkenroda. - Kali, verw. Salze u. Erdöl 26: 25-33, 39-43. Halle (Saale).

ALDRICH, L.T., NIER, A.O. (1948): Argon 40 in potassium minerals. - Phys. Review 74: 876-877. Lancaster, Pennsylvania.

ALEXANDER, E.C. JR. (1974): Helium. - In: Handbook of Geochemistry, Vol. II: 2-I; ed. by K.H. WEDEPOHL, Springer, Berlin-Heidelberg-New York.

ANSORGE, D.W. (1987): Eine gesalzene Lösung. Kavernendeponie an der Nordseeküste. - Umweltmagazin 16: 56-58. Vogel.

ANTONOW, P.L., GLADYSCHEWA, G.A., KOSLOW, W.P. (1958): Die Diffusion von Kohlenwasserstoffgasen durch Steinsalz. - Z. angew. Geol. 4: 387-388. Berlin.

BAAR, A. (1952): Grubengase im Südharz-Kalibergbau. - Bergbautechnik 2: 469-473. Berlin.

BAAR, A. (1954): Zur Schlagwetterbekämpfung im Südharz-Kalibergbau. - Bergbautechnik 4: 339-343. Berlin.

BAAR, A. (1958): Über gleichartige Gebirgsverformungen durch bergmännischen Abbau von Kaliflözen bzw. durch chemische Umbildung von Kaliflözen in geologischer Vergangenheit. - Freiberger Forschungshefte, A 123: 137-159. Berlin.

BAAR, C.A. (1977): Applied Salt-Rock Mechanics 1. The in-situ behavior of salt rocks. - pp. 294, Elsevier Sci. Pub. Amsterdam-Oxford-New York.

BARDET, G. (1976): Experience de sept années de stockage de déchets radioactifs solides de faible et moyenne activité en surface ou en tranchées bétonnées. - In: Proc. Symposium on the Management of Radioactive Wastes from the Nuclear Fuel Cycle, Vienna, March 22-26, 1976; Vol. II: 351-357. IAEA, Vienna.

BAUER, G. (1991): Kryogene Klüfte in norddeutschen Salzdiapiren? - Zbl. Geol. Paläont., Vol. 4, Part 1: 1247-1261. Hannover.

BAUMERT, B. (1928): Über Laugen- und Wasserzuflüsse im deutschen Kalibergbau. - pp. 90, Dissertation, TH Aachen. Gebr. Gerstenberg, Hildesheim.

BAUMERT, B. (1952): Die Wasser- und Laugengefahr im deutschen Kalibergbau. - pp. 212, unpub.

BECK, K. (1912): Petrographisch-geologische Untersuchung des Salzgebirges im Werra-Fulda-Gebiet der deutschen Kalisalzlagerstätten. - Z. prak. Geol. 20: 133-158. Berlin.

BELCHIC, H.C. (1960): The Winnfield salt dome, Winn Parish, Louisiana. - In: Interior salt domes and tertiary stratigraphy of North Louisiana, Guide Book: 29-47. 1960 Spring Field Trip, Shreveport Geological Society.

BIELER, H., CLAUS, F. (1988): Untertagedeponien als Entsorgungszentren. - Raumplanung 40: 50-55.

BILTZ, W., MARCUS, E. (1909): Über das Vorkommen von Ammoniak und Nitrat in den Kalisalzlagerstätten. - Z. anorg. Chem. 62: 183-202; 64: 215-216. Hamburg-Leipzig.

BOEKE, H.E. (1909): Eine neue Verbindung von Eisenchlorür und Chlormagnesium. - Kali III: 147. Halle (Saale).

BOEKE, H.E. (1911): Über die Eisensalze in den Kalisalzlagerstätten. - N. Jb. Min. Geol. Pal. I: 48-76. Stuttgart.

BOL'SHAKOV, YU.YA. (1972): Gas accumulation in the evaporite formation of the Solikamsk basin. - Dokl. Acad. Sci. USSR, Earth Sci. Sect., 204: 213-214. U.S.A.

BORCHERT, H. (1940): Salzlagerstätten des deutschen Zechsteins. - pp. 196, Archiv für Lagerstättenforschung 67. Berlin.

BORCHERT, H. (1959): Ozeane Salzlagerstätten. - pp. 237, Gebrüder Borntraeger. Berlin.

BORCHERT, H., MUIR, R.O. (1964): Salt deposits. The origin, metamorphism and deformation of evaporites. - pp. 338, D. van Nostrand. London.

BORN, H.-J. (1934/35): Der Bleigehalt der Norddeutschen Salzlager und seine Beziehungen zu radioaktiven Fragen. - Chemie der Erde 9: 66-87. Jena.

BORN, H.-J. (1936): Geochemische Zusammenhänge zwischen Helium-, Blei- und Radiumvorkommen in deutschen Salzlagerstätten. - Kali, verw. Salze und Erdöl 30: 41-45. Halle (Saale).

BORN, H.-J. (1959): Zur Frage der geochemischen Folgerungen aus den Hahnschen Arbeiten über Mitfällungen. - In: Beiträge zur Physik und Chemie des 20. Jahrhunderts, ed. by O.R. FRISCH, F.A. PANETH, F. LAVES, P. ROSBAUD, 130-134. Vieweg & Sohn, Braunschweig.

BORNEMANN, O. (1991): Zur Geologie des Salzstocks Gorleben nach den Bohrergebnissen. pp. 67, BfS Schriften 4/91. Salzgitter.

BORNEMANN, O., FISCHBECK, R. (1986): Ablaugung und Hutgesteinsbildung am Salzstock Gorleben. - Z. dt. geol. Ges. 137: 71-83.

BORNEMANN, O., FISCHBECK, R. (1989): Quarternary subrosion and transformation of Stassfurt potash seam (Zechstein 2) at top of Gorleben salt dome, Federal Republic of Germany. - Am. Ass. Petrol. Geol. Convention 1989, abstracts, p. 19.

BORNEMANN, O., FISCHBECK, R., ZIRNGAST, M. (1989): Verkarstete Anhydritgesteine im Flankenbereich des Salzstocks Gorleben - Reste einer früh- oder präkretazischen Hutgesteinsbildung? - DGG Jahrestagung, Braunschweig, Oktober 1989.

BORNEMANN, O., JARITZ, W., WITTROCK, J. (1988): Geotechnische Erkundung und Standsicherheitskriterien Bergwerk Gorleben. Teilprojekt I: Geologisches und geotechnisches Untersuchungsprogramm. - BGR-Forschungsvorhaben KWA 85049. Hannover.

BORSTEL, L.E. VON (1991): Die Charakterisierung des Stoffbestandes von fluid inclusions in Zechsteinevaporiten mittels der Lösungsgleichgewichte mariner Salzsysteme. - Kali u. Steinsalz 10: 409-416. Essen.

BORSTEL, L.E. VON (1992): Lösungen in marinen Zechsteinevaporiten Deutschlands. - pp. 324, Dissertation, Technische Universität Clausthal.

BRAITSCH, O. (1960): Mineralparagenesis und Petrologie der Staßfurtsalze in Reyershausen. - Kali und Steinsalz 3: 1-14.

BRAITSCH, O. (1962): Entstehung und Stoffbestand der Salzlagerstätten. - pp. 232, Springer, Berlin-Göttingen-Heidelberg.

BRAITSCH, O. (1971): Salt deposits. Their origin and composition. - pp. 297, Springer, Berlin-Heidelberg-New York.

BRAITSCH, O., HERRMANN, A.G. (1962): Zur Bromverteilung in salinaren Salzsystemen bei 25 °C. - Naturwissenschaften 15: 346-348.

BRAITSCH, O., HERRMANN, A.G. (1963): Zur Geochemie des Broms in salinaren Sedimenten. Teil I: Experimentelle Bestimmung der Br-Verteilung in verschiedenen natürlichen Salzsystemen. - Geochim. Cosmochim. Acta 27: 361-391.

BRENNECKE, P., SCHUMACHER, J. (1990): Anfall radioaktiver Abfälle in der Bundesrepublik Deutschland - Abfallerhebung für das Jahr 1989. - BfS, ET 1/90. Salzgitter.

BRUMSACK, H.J., HEINRICHS, H. (1984): Potentielle Emissionen einer Müllverbrennungsanlage. - In: Wohin mit dem Müll?, 25-32, Arbeitskreis Ökologie. Göttingen.

BRUMSACK, H.J., HEINRICHS, H., LANGE, H. (1984): West German coal power plants as sources of potentially toxic emissions. - Environ. Tech. Lett. 5: 7-22.

BURGBACHER, G., ROTH, K., STERZ, O. (1986): Planfeststellung für die Deponierung der Rückstände aus Rauchgasreinigungsanlagen von Müllheizkraftwerken im Steinsalzbergwerk Heilbronn. - TÜV Stuttgart, Abschlußbericht vom 17.09.1986.

Cadelli, N., Cottone, G. (1990): PAGIS - Performance assessment of geological isolation systems for radioactive waste. - Sitzungsbericht des PAGIS-Informationstags vom 30. Juni 1989, Comission of the European Communities. Luxembourg, (in German).

Cadelli, N., Cottone, G., Orlowski, S., Bertozzi, G., Girardi, F., Saltelli, A. (1988): PAGIS - Performance assessment of geological isolation systems for radioactive waste. Summary - Comission of the European Communities. Luxembourg.

CHAPMAN, N.A., McKINLEY, I.G., HILL, M.D. (1987): The geological disposal of nuclear waste. - pp. 280, John Wiley & Sons, Chichester-New York-Brisbane-Toronto-Singapore.

CHEREPENNIKOV, cited after SAVCHENKO (1958).

DAS, N.; HORITA J.; HOLLAND, H. D. (1989): Chemistry of fluid inclusions in halite from the Salina Group of Michigan Basin: Implications for Late Silurian seawater and the origin of sedimentary brines. - Geochim. Cosmochim. Acta. 54: 319-327.

D'ANS, J. (1933): Die Lösungsgleichgewichte der Systeme der Salze ozeanischer Salzablagerungen. - pp. 254, Verlagsges. f. Ackerbau, Berlin.

D'ANS, J., FREUND, H.E. (1954): Versuche zur geochemischen Rinneitbildung. - Kali u. Steinsalz 1, 6: 3-9. Essen.

DEECKE, H. (1949): Schichtenfolge und Tektonik des Rotliegend-Zechstein-Salzaufbruchs von Heide. - In: Erdöl und Tektonik in Nordwestdeutschland, ed. by BENTZ, A.: 173-190. Hannover-Celle.

DISPOSAL OF HIGH-LEVEL WASTE FROM NUCLEAR POWER PLANTS IN DENMARK. SALT DOME INVESTIGATIONS. Vol. II, Geology, text (1981). - Report prepared by Elsam and Elkraft, p. 93, 104, 109. Ballerup und Fredericia.

DUCHROW, G. (1959): Untersuchungen in Kohlensäure-Rachelfeldern der Grube Menzengraben. - Bergakademie 9: 586-594. Berlin.

DUDYREV, A.N., SUNGUROVA, Z.N. (1963): Gaszusammensetzung und Charakter der Ausgasungen in der Kaligrube Solikamsk. - cited after GIMM and ECKART (1968).

ECKART, D., GIMM, W., THOMA, K. (1966): Plötzliche Ausbrüche von Gestein und Gas im Bergbau. -pp. 235, Freiberger Forschungshefte, A 409. Leipzig.

EHRHARDT, K. (1980): Exploration eines neuen Baufeldes im Grubenbetrieb des Steinsalzbergwerks Braunschweig-Lüneburg der Kali und Salz AG. - In: Fifth Symposium on Salz, Vol. I: 231-238, ed. by A. H. COOGAN and L. HAUBER, The Northern Ohio Geological Society. Cleveland, Ohio.

EHRLICH, D., RÖTHEMEYER, H., STIER-FRIEDLAND, G., THOMAUSKE, B. (1986): Langzeitsicherheit von Endlagern. Zeitrahmen für Sicherheitsbetrachtungen - Bewertung der Subrosion des Salzstocks Gorleben. - Atomwirtschaft-Atomtechnik 31: 231-236. Identical text in: PTB informiert, 1987, 1: 2-7.

ELERT, K.H., FREUND, W. (1969): Untersuchungen zum Auftreten von feindispers verteilten Kohlenwasserstoffgasen im Kaliflöz "Staßfurt". - Bergakademie 21: 584-589. Berlin.

ELERT, K.-H., HENNING, I. (1988): Bitumen A in Gesteinen des Zechsteins. - Angew. Geol. 34: 139-144. Berlin.

ELERT, K.-H., HENNING, I., KNABE, H.-J. (1988): Untertägige Erdölvorkommen und ihre bergbausicherheitliche Beurteilung. - Angew. Geol. 34: 71-76. Berlin.

ELERT, K.H., KNABE, H.-J. (1982): Gazovye vkljucenija v porodach cechstejna 2 i 3 (GDR). - In: Nefte-Gazonost regionov drevnego solenakoplenija: 146-153. Izdatelstvo nauka, Sibirskoe Otdelenia, Novosibirsk, (in Russian).

ELSAM AND ELKRAFT REPORT (1981): Disposal of high-level waste from nuclear power plants in Denmark. - Vol. I-V. Elsam, Ballerup, Denmark.

ERDA (1977a): Waste management operations. - Idaho National Engineering Laboratory, Idaho, September 1977, Erda 1536.

ERDA (1977b): Waste management operations. - Savannah River Plant, Aiken, South Carolina, September 1977, Erda 1537.

ERDMANN, E. (1910): Zwei neuere Gasausströmungen in deutschen Kalisalzlagerstätten. - Kali IV: 137-142. Halle (Saale).

ERLER, H. (1957): Vorkommen von Erdgas und Erdöl in den Kaliwerken und Bohrungen nördlich vom Harz. - Bergakademie Freiberg, unpub., cited after J. LÖFFLER and G. SCHULZE (1962).

FABRICIUS (1984): Studies of fluid inclusions in halite and euhedral quartz crystals from salt domes in the Norwegian-Danish basin.- Geol. Surv. Denmark, Salt Research Projekt EFP-81: 7-31.

FIEGE, K. (1934): Übersicht über das Vorkommen der Erdöle, Erdgase und Asphalte in Deutschland. - Kali, verw. Salze und Erdöl 28: 183-191, 200-204, 215-219, 228-230, 242-245. Halle (Saale).

FINKENWIRTH, A., JOHNSSON, G. (1980): Die Untertage-Deponie Herfa-Neurode bei Heringen/Werra. - In: Fifth Symposium on Salt, Vol. 1: 239-249. Ed. by A.H. COOGAN and L. HAUBER, The Northern Ohio Geological Society. Cleveland, Ohio.

FISCHBECK, R. (1984): Umwandlungen von Kalisalzgestein aus der Ronnenberg-Gruppe (z3) im Salzstock Gorleben. - Kali u. Steinsalz 9: 61-65.

FISCHBECK, R., BORNEMANN, O. (1988): Genetische Überlegungen aufgrund von Brom-Bestimmungen an halitischen Kluftfüllungen in Salzgesteinen des Salzstocks Gorleben, Niedersachsen. - Fortsch. Min. 66, Beih. 1: 35. Stuttgart.

FÖRSTER, S. (1974): Durchlässigkeits- und Rißbildungsuntersuchungen zum Nachweis der Dichtheit von Salzkavernen. - Neue Bergbautechnik 4: 278-283. Berlin.

FÖRSTER, S. (1985): Gasdruckbelastbarkeit und Rißbildung der für die unterirdische Gasspeicherung in Kavernen bedeutsamen Salinargesteinen des Zechsteins. - pp. 114, Freiberger Forschungshefte, A 724, Leipzig.

FRANTZEN, W. (1894): Bericht über neue Erfahrungen beim Kalibergbau in der Umgebung des Thüringer Waldes. - Jb. d. Königlich Preußischen Geol. Landesanstalt und Bergakademie XV: LX-LXI. Berlin.

FREYER, H.D. (1973): Nachweis atmosphärischer Gase in gasarmen Salzgesteinen. - Kali u. Steinsalz 6: 117-121. Essen.

FREYER, H.D. (1978): Degradation products of organic matter in evaporites containing trapped atmospheric gases. - Chem. Geol. 23: 293-307. Amsterdam.

FREYER, H.D., WAGENER, K. (1975): Review on present results on fossil atmospheric gases trapped in evaporites. - Pure and applied geophysics 113: 403-418. Basel.

FÜRER, G. (1989): Entsorgung von Fremdabfällen in bergbauähnlichen Hohlräumen - Gesetzliche und bergtechnische Rahmenbedingungen. - Erzmetall 42: 189-196. Weinheim.

FYFE, W.F., BABUSKA, V., PRICE, N.J., SCHMID, E., TSANG, C.F., UYEDA, S., VELDE, B. (1984): The geology of nuclear waste disposal. - Nature 310: 537-540.

GENTNER, W., GOEBEL, K., PRÄG, R. (1954): Argonbestimmungen an Kalium-Mineralien. III. Vergleichende Messungen nach der Kalium-Argon- und Uran-Helium-Methode. - Geochim. Cosmochim. Acta, 5: 124-133. London.

GENTNER, W., PRÄG, R., SMITS, F. (1953): Argonbestimmungen an Kalium-Mineralien. II. Das Alter eines Kalilagers im Unteren Oligozän. - Geochim. Cosmochim. Acta, 4: 11-20. London.

GERLING, P., BEER, W., BORNEMANN, D. (1991): Gasförmige Kohlenwasserstoffe in Evaporiten des deutschen Zechsteins. - Kali u. Steinsalz 10: 376-383. Essen

GIESEL, W. (1968): Kohlensäureausbrüche im Kalibergbau an der Werra - Grundlagen und Prognosemöglichkeiten. - Kali u. Steinsalz 5: 103-108. Essen.

GIESEL, R.-J., HAASE, G., MARKGRAF, P., SALZER, K., THOMA, K. (1989): Drei Jahrzehnte Ausbruchsforschung im Kalibergbau des Werrareviers der DDR. - Z. geol. Wiss. 17: 336-346. Berlin.

GIMM, W. (1954): Kohlensäure und Kohlenwasserstoffgase im Kalibergbau der DDR und Methoden zur Bekämpfung der Gasgefahren. - Bergbautechnik 4: 587-592, 656-662. Berlin.

GIMM, W. (1955): Salzgebundene Gase im Kalibergbau. - Freiberger Forschungshefte, A 42: 103-133. Leipzig.

GIMM, W. (1964): Überblick über die von der Forschungsgemeinschaft "Mineralgebundene Gase" bearbeiteten Probleme sowie Ergebnisse der Forschungsarbeiten. - Freiberger Forschungshefte, A 304: 5-49. Leipzig.

GIMM, W., ECKART, D. (1968): Vorkommen und Bekämpfung natürlicher Gase. - In: Kali und Steinsalzbergbau, Vol. 1: 544-594. Ed. by W. GIMM, VEB Deutscher Verlag für Grundstoffindustrie. Leipzig.

GOLUBIC, S., KRUMBEIN, W., SCHNEIDER, J. (1979): The carbon cycle. - In: Biochemical cycling of mineralforming elements. Eds.: P.A. TRUDINGER, D.J. SWAINE; 29-45. Elsevier Sci. Pub., Amsterdam-Oxford-New York.

GOMM, H. (1982): Literaturstudie über Kohlenwasserstoff- und Kondensateinschlüsse in Salzvorkommen. - pp. 30, unpub., Kavernen Bau- und Betriebs-GmbH, Hannover.

GORDON, W.A. (1975): Distribution by latitude of Phanerozoic evaporite deposits. - J. Geology 83: 671-684. Chicago.

GOVERNMENT RELEASE (1989): Weltweit beachtete untertägige Sonderabfalldeponie in Niedersachsen. - Der Bundesminister für Forschung und Technologie, 5. Juli 1989. Bonn.

GROPP (1919): Gasvorkommen in Kalisalzbergwerken in den Jahren 1907 - 1917. - Kali 13: 33-42, 70-76. Halle (Saale).

GRÜBLER, G. (1983): Kohlenwasserstoffe im Salzstock. - In: Zusammenfassender Zwischenbericht über bisherige Ergebnisse der Standortuntersuchung in Gorleben. - 35-38. Physikalisch-Technische-Bundesanstalt, Braunschweig.

GRÜBLER, G. (1984a): Gasvorkommen in den Schachtvorbohrungen Go 5001 und Go 5002. - In: Entsorgung, Vol. 3: 165-181. Bericht von einer Informationsveranstaltung des Bundes vor dem Schachtabteufen, Salzstock Gorleben. Ed.: Bundesministerium für Forschung und Technologie. Bonn.

GRÜBLER, G. (1984b): Auslegung von Erkundungs-Schächten bei Gaszutritten in den Schachtvorbohrungen. - In: Entsorgung, Vol. 3; 427-439. Bericht von einer Informationsveranstaltung des Bundes vor dem Schachtabteufen, Salzstock Gorleben. Ed.: Bundesministerium für Forschung und Technologie. Bonn.

GRÜBLER, G., REPPERT, D. (1983): Bericht über die in den Schachtvorbohrungen Go 5001 und Go 5002 angetroffenen KW-Kondensate/-Gase und deren Untersuchungsergebnisse. - pp. 53, Physikalisch-Technische Bundesanstalt Braunschweig, unpub.

GUDOWIUS, E. (1984): Neue Auswertung bekannter und unveröffentlichter Untersuchungen am Kali-Forschungsinstitut der Kali und Salz AG, Hannover, unpub.

GÜNTHER, cited after GIMM (1964).

GUTSCHE, A. (1988): Mineralreaktionen, Stofftransporte und Stoffbilanzen im Kontaktbereich Basalt-Sylvinit am Beispiel 1. Begleitflöz im Hangenden des Kalisalzflözes Hessen (K1H), Kaliwerk Hattorf. - Diplomarbeit. Göttingen, unpub.

GUTSCHE, A., HERRMANN, A.G. (1988): Wechselwirkungen zwischen fluiden Phasen und Evaporiten im Nahbereich von Basaltgängen. - Fortschr. Mineral. 66, Beih. 1: 49.

HAHN, O. (1932a): Über Blei und Helium in ozeanischen Alkalihalogeniden. - Naturwissenschaften 20: 86-87. Berlin.

HAHN, O. (1932b): Radioaktivität und ihre Bedeutung für Fragen der Geochemie. - Sitzungsberichte d. Preuß. Akad. d. Wiss., Phys.-Math. Kl.: 2-14. Berlin.

HALBOUTY, M.T. (1979): Salt domes. Gulf Region, United States & Mexico. - Second Edition, pp. 561, Gulf Publishing Company. Houston-London-Paris-Tokyo.

HARDY, H.R. JR., MANGOLDS, A. (1980): Investigation of residual stresses in salt. - In: Fifth Symposium on Salt, Vol. I: 55-63. Ed. by A.H. COOGAN and L. HAUBER. The Northern Ohio Geological Society. Cleveland, Ohio.

HARTECK, P., SUESS, H. (1947): Der Argongehalt kalihaltiger Minerale und die Frage des dualen Zerfalls von $K^{40}$. - Die Naturwissenschaften 34: 214-215. Berlin-Göttingen.

HARTWIG, G. (1954): Zur Kohlensäureführung der Werra- und Fulda-Kalisalzlager. - Kali u. Steinsalz 1: 3-26. Essen.

HARVIE, C. E., WEARE, J. H. (1980): The prediction of mineral solubilities in natural waters: the Na-K-Mg-Ca-Cl-$SO_4$-$H_2O$ system from zero to high concentration at 25 °C. - Geochim. Cosmochim. Acta, 44: 981-997.

HEIDORN, W. (1926): Stratigraphie und Tektonik des Hope-Lindwedeler Zechsteinsalzstocks. - Jb. nieders. geol. Ver. 19: 8-42. Hannover.

HEINRICHS, H., BRUMSACK, H.J., LANGE, H. (1984): Emissionen von Steinkohle- und Braunkohlekraftwerken in der Bundesrepublik Deutschland. - Fortschr. Miner. 62: 79-105.

HEMMANN, M. (1989): Gasvorkommen in der Steinsalzgrube Bernburg-Gröna. — Z. geol. Wiss. 17: 389-400. Berlin.

HEMPEL, D. (1974): Neuere Erkenntnisse über das Auftreten natürlicher brennbarer Gase in Kaligruben der DDR. - Neue Bergbautechnik 4: 592-596. Berlin.

HEMPEL, D. (1989): Umfang und Gestaltung des Schlagwetterschutzes im Kali- und Kupferschieferbergbau. - Z. geol. Wiss. 17: 419-429. Berlin.

HEMPEL, D., HAISLER, K., WOLFF, H., RAUER, H. (1981): Neue Erkenntnisse über Umfang und Gestaltung des Schlagwetterschutzes im Kali- und Steinsalzbergbau der DDR. - Neue Bergbautechnik 11: 636-639. Leipzig.

HENNIES, J. (1989): Einlagerungsmöglichkeiten im Bergbau. - Vortrag, Vorstandssitzung der Wirtschaftsvereinigung Bergbau e.V., 07.03.1989, unpub.

HERRMANN, A.G. (1961a): Über das Vorkommen einiger Spurenelemente in Salzlösungen aus dem deutschen Zechstein. - Kali u. Steinsalz 3: 209-220. Essen.

HERRMANN, A.G. (1961b): Über die Einwirkung Cu-, Sn-, Pb und Mn-haltiger Erdölwässer auf die Staßfurt-Serie des Südharzbezirkes. - N. Jb. Miner., Mh., 60-67. Stuttgart.

HERRMANN, A.G. (1964): Geochemische Untersuchungen an einem Vorkommen von Fasersteinsalz. - Beitr. Miner. Petrol. 10: 374-378.

HERRMANN, A. G. (1972): Bromine distribution coefficients for halite precipitated from modern sea water under natural conditions. - Contrib. Miner. Petrol. 37: 249-252.

HERRMANN, A.G. (1979): Geowissenschaftliche Probleme bei der Endlagerung radioaktiver Substanzen in Salzdiapiren Norddeutschlands. - Geol. Rdsch. 68: 1076-1106.

HERRMANN, A.G. (1980): Geochemische Prozesse in marinen Salzablagerungen: Bedeutung und Konsequenzen für die Endlagerung radioaktiver Substanzen in Salzdiapiren . - Z. dt. Geol. Ges. 131: 433-559.

HERRMANN, A.G. (1982): Probenahme von Salzlösungen in Kali- und Steinsalzbergwerken. - Kali u. Steinsalz 8: 237-242. Essen.

HERRMANN, A.G. (1983a): Radioaktive Abfälle. Probleme und Verantwortung. - pp. 256, Springer. Berlin-Heidelberg-New York.

HERRMANN, A.G. (1983b): Lösungen. - In: Zusammenfassender Zwischenbericht über bisherige Ergebnisse der Standortuntersuchung in Gorleben. - 38-46. Physikalisch-Technisch-Bundesanstalt. Braunschweig.

Herrmann, A.G. (1984a): Die Entstehung und Herkunft von Lösungen im Salzstock Gorleben. - In: Entsorgung, Vol. 3: 441-451. Bericht von einer Informationsveranstaltung des Bundes vor dem Schachtabteufen, Salzstock Gorleben. Ed.: Bundesministerium für Forschung und Technologie. Bonn.

HERRMANN, A.G. (1984b): Forschungsprojekt "Geochemische Untersuchungen an gasführenden Evaporitgesteinen der Werra-Serie des Zechsteins". - Geochemisches Institut der Universität Göttingen, 10.12.1984, unpub.

HERRMANN, A.G. (1986): Forschungsprojekt "Geochemische Untersuchungen an gasführenden Evaporitgesteinen der Zechstein-Zyklen 1-4". - Geochemisches Institut der Universität Göttingen, 14.10.1986, unpub.

HERRMANN, A.G. (1987a): Untergrund-Deponie anthropogener Schadstoffe. - Fortschr. Miner. 65: 307-323. Stuttgart.

HERRMANN, A.G. (1987b): Forschungskonzept fluid inclusions. - Geochemisches Institut der Universität Göttingen, unpub.

HERRMANN, A.G. (1988a): Die Untergrund-Deponie anthropogener Schadstoffe. - In: Die Erde. Dynamische Entwicklung, menschliche Eingriffe, globale Risiken. Ed.: K. GERMANN, G. WARNECKE, M. HUCH; p. 183-200. Springer, Berlin-Heidelberg-New York-London-Paris-Tokyo.

HERRMANN, A.G. (1988b): Die Untergrund-Deponie chemischer und radioaktiver Schadstoffe. Versuch einer quantitativen Beurteilung aus der Sicht der Geowissenschaften. - In: Vol. 3: 353-389. Pilotprojekt zu Clausthaler Kursen zur Umwelttechnik. 30.05. - 01.06.1988 in Goslar. CUTEC, Forschungsverbund Umwelttechnik der TU Clausthal.

HERRMANN, A.G. (1988c): Geowissenschaftliche Grundlagen für die langfristig sichere Deponie chemischer und radioaktiver Schadstoffe. - In: Vol. 7: 27-66. Deponieren von Abfällen. Clausthaler Kursus zur Umwelttechnik. 17.10. - 20.10.1988 in St. Andreasberg. CUTEC, Forschungsverbund Umwelttechnik der TU Clausthal.

HERRMANN, A.G. (1988d): Gase in marinen Evaporiten. - pp. 33, PTB informiert 2/88.

HERRMANN, A.G. (1989): Natürliche Stoffkreisläufe und die Deponie anthropogener Abfälle. - Naturwissenschaftliche und historische Beiträge zu einer ökologischen Grundbildung. - 114-125, Eds. B. HERRMANN and A. RÜDDE. Das Niedersächsische Umweltministerium. Hannover.

HERRMANN A.G. (1991a): Die Untergrund-Deponie anthropogener Abfälle in marinen Evaporiten. - Materialien zur Umweltforschung, Heft 18: pp. 109, Ed.: Rat von Sachverständigen für Umweltfragen. Metzler-Poeschel, Stuttgart.

HERRMANN, A.G.(1991b): Dynamische Prozesse in der Natur als Kriterien für die langfristig sichere Deponierung anthropogener Abfälle. - In: Tatort Erde, Eds.: G.WARNECKE, M. HUCH, K. GERMANN: 86-111. Springer, Berlin-Heidelberg-New York-London-Paris-Tokyo-Hong Kong-Barcelona-Budapest.

HERRMANN, A. G. (1991c): Fraktionierung im Stoffbestand der Zechsteinevaporite Mittel- und Norddeutschlands. - Zbl. Geol. Paläont. Vol.4, Part 1: 1091-1106.

HERRMANN, A. G. (1992): Endlager für radioaktive Abfälle Morsleben (ERAM). Lösungszuflüsse in den Grubenfeldern Marie und Bartensleben: Stoffbestand, Herkunft, Entstehung. - pp. 480, BfS Schriften. Salzgitter.

HERRMANN, A. G., BORSTEL, L. E. VON (1991): The composition and origin of fluid inclusions in Zechstein evaporites of Germany. - N.Jb. Miner. Mh.: 236-269.

HERRMANN, A.G., BRUMSACK, H.J., HEINRICHS, H. (1985): Notwendigkeit, Möglichkeiten und Grenzen der Untergrund-Deponie anthropogener Schadstoffe. - Naturwissenschaften 72: 408-418.

HERRMANN, A.G., KNAKE, D., SCHNEIDER, J., PETERS, H. (1973): Geochemistry of modern seawater and brines from salt pans: Main components and bromine distribution. - Contr. Miner. Petr. 40: 1-24.

HERRMANN, A.G., KNIPPING, B. (1988a): Stoffbestand und Langzeitsicherheit von Schadstoffdeponien in Salzstöcken. - In: PTB informiert, 1/88: 21-30. Braunschweig.

HERRMANN, A.G., KNIPPING, B. (1988b): Stoffbestand und Langzeitsicherheit von Schadstoff-Deponien in Salzstöcken. - In: Vol. 7: 349-376. Clausthaler Kursus zur Umwelttechnik. 17.10. - 20.10. 1988 in St. Andreasberg. CUTEC, Forschungsverbund Umwelttechnik der TU Clausthal.

HERRMANN, A.G., KNIPPING, B. (1989): Stoffbestand von Salzstöcken und Langzeitsicherheit für Endlager radioaktiver Abfälle. - In: PTB informiert, 1/89: pp. 50. Braunschweig.

HERRMANN, A. G., KNIPPING, B., SCHRÖDER, K., BORSTEL, L. E. VON (1990): EFI: Vorrichtung zur Extraktion von fluid inclusions in Salzmineralen. - E. J. Miner., Bh. 1: 99.

HERRMANN, A. G., KNIPPING, B., SCHRÖDER, K., BORSTEL, L. E. VON (1991): The quantitative analysis of fluid inclusions in marine evaporites. - N. Jb. Min. Mh.: 39-48.

HERRMANN, A. G., SIEBRASSE, G., KÖNNECKE, K. (1978): Computerprogramme zur Berechnung von Mineral- und Gesteinsumbildungen bei der Einwirkung von Lösungen auf Kali- und Steinsalzlagerstätten (Lösungsmetamorphose). - Kali u. Steinsalz 7: 288-299. Essen.

HEYMEL, cited after GIMM (1954).

HOEFS, J. (1969): Carbon. - In: Handbook of Geochemistry, Vol. II-1, 6-K, ed. by K.H. WEDEPOHL. Springer, Berlin-Heidelberg-New York.

HOFFMANN (1963), cited after personal communication by R. SCHRADER (1964).

HOFFMANN, R.O. (1961): Die Mineralzusammensetzung der in Wasser schwer löslichen Rückstände von Filterschlämmen und Rohsalzen einiger mitteldeutscher Kaliwerke. - Bergakademie 13: 237-248.

HOFRICHTER, E. (1976): Zur Frage der Porosität und Permeabilität von Salzgesteinen. - Erdoel-Erdgas-Ztschr. 92: 77-80. Hamburg-Wien.

HOFRICHTER, E. (1980a): Probleme der Endlagerung radioaktiver Abfälle in Salzformationen. - Z. dt. geol. Ges. 131: 409-430. Hannover.

HOFRICHTER, E. (1980b): Die Untertage-Deponie im ehemaligen Kalisalzbergwerk Thiederhall. - In: Müll- und Abfallbeseitigung, Vol. 5: 55. Lfg., I/80, 1-4, Eds.: KUMPF, MAAS, STRAUB.

HOLSER, W. T. (1963): Chemistry of brine inclusions in Permian Salt from Hutchinson, Kansas. - In: First Symp. on Salt. The Northern Ohio Geological Society. Cleveland, Ohio.

HOPPE, W. (1960): Die Kali- und Steinsalzlagerstätten des Zechsteins in der Deutschen Demokratischen Republik. Teil 1: Das Werra-Gebiet. - Freiberger Forschungshefte, C 97/I, pp. 166. Berlin.

HORITA, J.; FRIEDMANN, T.J.; LAZAR, B.; HOLLAND, H.D. (1991): The composition of Permian seawater. - Geochim. Cosmochim. Acta 55: 417-432.

HORITA, J., MATSUO, S. (1986): Extraction and isotopic analysis of fluid inclusions in halites. - Geochem. Journal 20: 261-272.

HORSTMANN, U., V. STRUENSEE, G. (1981): Der Salzstock Hope. - In: Erläuterungen zu Blatt Nr. 3324 Lindwedel: 57-58. Geologische Karte von Niedersachsen 1:25 000. Niedersächsisches Landesamt für Bodenforschung, Hannover.

HOY, R.B., FOOSE, R.M. & O'NEILL, B.J. JR. (1962): Structure of Winnfield salt dome, Winn Parish, Louisiana. - Bull. Amer. Assoc. Petrol. Geol. 46: 1444-1459.

HUNER, J. JR. (1939): Geology of Caldwell and Winn Parishes. - Geological Bulletin No. 15, pp. 356, Department of Conservation, Louisiana Geological Survey. New Orleans.

HUNT, J.M. (1979): Petroleum Geochemistry and Geology. - pp. 617, W.H. Freeman and Company. San Francisco.

HYMAN, D.M. (1982): Methodology for determining occluded gas contents in domal rock salt. - Bureau of Mines Report of Investigations 8700, United States Department of Interior. Pittsburgh, Pennsylvania.

IANNACCHIONE, A.T., GRAU, R.H. III., SAINATO, A., KOHLER, T.M., SCHATZEL, S.J. (1984): Assessment of methans hazards in an anomalous zone of a Gulf Coast salt dome. - Bureau of Mines Report of Investigations 8861, United States Department of Interior. Pittsburgh, Pennsylvania.

IVANOV, A.A. (1935): Karnallity Verchnekamskoga mestorozdenija. - In: Solikamskie Karnallity: 5-12. Ed. by N.S. KURNAKOV, V.E. CIFRINOVIC, C.I. VOL'FKOVIC, Glavnaja Redakcija Gorno-Toplivnoj Literatury, Moskva-Leningrad, (in Russian).

IWS-SCHRIFTENREIHE 1 (1987): Symposium "Die Deponie - ein Bauwerk?". - pp. 413, Ed.: Trägerverein des Instituts für wassergefährdende Stoffe e.V. an der Technischen Universität Berlin.

JAHNE, H., PIELERT, P. (1964): Beitrag zur Ausbildung des hangenden Begleitflözes zum Kaliflöz "Hessen" im Werrakaligebiet unter besonderer Berücksichtigung der Schachtanlage "Marx-Engels" Unterbreizbach. - Ber. d. Geol. Ges. DDR 9: 641-665. Berlin.

JARITZ, W. (1973): Zur Entstehung der Salzstrukturen Nordwestdeutschlands. - Geol. J., Reihe A, Heft 10, pp. 77, E. Schweizerbart'sche Verlagsbuchhandlung. Stuttgart.

JARITZ, W. (1980): Einige Aspekte der Entwicklungsgeschichte der nordwestlichen Salzstöcke. - Z. dt. geol. Ges., 131: 387-408.

JENKS, G.H. (1972): Radiolysis and hydrolysis in salt-mine-brines. - pp. 99, ORNL-TM-3717, Oak Ridge National Laboratory. Oak Ridge, Tennessee.

JENKS, G.H., BOPP, C.D. (1977): Storage and release of radiation energy in salt in radioactive waste repositories. - pp. 102, ORNL-5058, Oak Ridge National Laboratory. Oak Ridge, Tennessee.

JENKS, G.H., SONDER, E., BOPP, C.D., WALTON, J.R., LINDENBAUM, S. (1975): Reaction products and stored energy released from irradiated sodium chloride by dissolution and by heating. - J. Phys. Chem. 79: 871-875.

JEWENTOW, J.S., MILESCHINA, A.G., KOMISSAROWA, I.N. (1973): Über die Erdöldurchlässigkeit fossiler Salze. - Z. angew. Geologie 19: 67-71. Berlin.

JEZIERSKI, H. (1989): Bergbau und Abfallwirtschaft - zukünftige Partner? - Erzmetall 42: 117-200. Weinheim.

JOCKWER, N. (1984): Laboratory investigations on radiolysis effects on rock salt with regard to the disposal of high-level radioactive wastes. - In: Mat. Rec. Soc. Symp. Proc. Vol. 26: 17-23, Elsevier Sci. Pub., Amsterdam.

JOCKWER, N. (1985): Basis and methods for in-situ measurements on liberation and generation of volatile components concerning in disposal of high-level radioactive waste in rock salt. - In: Design and Instrumentation of In-Situ Experiments in Underground Laboratories for Radioactive Waste Disposal. 352-358, ed. by B. COME, P. JOHNSTON, A. MÜLLER. Proceedings of a workshop jointly organized by the commission of the european communities & OECD nuclear Energy Agency/Brussels/15-17 May 1984, A.A. Balkema, Rotterdam-Boston.

JOCKWER, N., GROSS, S. (1985): Natural, thermal and radiolytical gas liberation in rock salt as a result of disposed high-level radioactive waste. - In: Mat. Res. Soc. Symp. Proc. Vol. 50: 587-594, Elsevier Sci. Pub., Amsterdam.

JOHNSEN, A. (1909a): Regelmäßige Verwachsung von Carnallit und Eisenglanz. - Zentralbl. Min. Geol. Pal.: 168-173. Stuttgart.

JOHNSEN, A. (1909b): Über die Entstehung von Wasserstoffgas in Kalisalzlagern. - Kali III: 118-119. Halle (Saale).

JOHNSSON, G. (1983): Underground disposal at Herfa-Neurode. - In: Hazardous Waste Disposal, Proceedings of the Symposium, 295-313, ed. by J.P.LEHMANN. Plenum Press, New York-London.

JOHNSSON, G. (1985): The operation of a waste disposal facility which accepts foreign hazardous waste. - In: Transfrontier Movements of Hazardous Wastes, 156-170, Paris: OECD.

JUNGHANS, R. (1955): Der schwere $CO_2$-Ausbruch auf der Schachtanlage Menzengraben des VEB Kaliwerk Heiligenroda am 7.7.1953, seine Ursachen und Folgen. - Bergbautechnik 3: 457-462, 579-589. Berlin.

KARLIK, B. (1939): Der Heliumgehalt von Steinsalz und Sylvin. - Mikrochemie, vereinigt mit Mikrochimica Acta 27: 216-230, Wien.

KÄSTNER, H. (1964): Beobachtungen zum Auftreten von Kohlenwasserstoffen im Werrakaligebiet. - Z. angew. Geol. 10: 359-365. Berlin.

KÄSTNER, H. (1968): Zur Entstehung und Verbreitung von Kohlendioxidlagerstätten. - Z. angew. Geol. 14: 316-323. Berlin.

KÄSTNER, H. (1969): Zur Geologie der Kalisalz- und Kohlensäurelagerstätten im südlichen Werra-Kaligebiet. -Abh. Zentr.Geol. Inst., Heft 11: pp. 96. Berlin.

KAUFMANN, D. W. (ED) (1968): Sodium chloride. The production and properties of salt and brine. - pp. 743, Am. Chem. Soc., Monograph series, facsimile of the 1960 ed., Hafner Pub. Co., New York.

KAUTSKY, H. (1982): Ausbreitungsvorgänge künstlicher Radionuklide im Meer. - Meerestechnik 13: 47-51, Düsseldorf.

KBB-Untertagespeicher (1984): - Druckschrift HSW 605/6.84 der Kavernen Bau- und Betriebs-GmbH, Hannover.

KIND, D. (1991): Vorwort. - In: Endlagerung radioaktiver Abfälle. Wegweiser für eine verantwortungsbewußte Entsorgung in der Industriegesellschaft. - pp. 275, ed. by H. RÖTHEMEYER, VCH Verlagsgesellschaft, Weinheim.

KINSMAN, D.J.J., BOARDMAN, M., BORCSIK, M. (1974): An experimental determination of the solubility of oxygen in marine brines. - Fourth Symposium on Salt, Vol. I: 325-327, ed. by A.H. COOGAN, The Northern Ohio Geological Society. Cleveland, Ohio.

KIRCHHEIMER, F. (1978): Das blaue Steinsalz. - Kali u. Steinsalz 7: 271-287. Essen.

KNABE, H.-J. (1989): Zur analytischen Bestimmung und geochemischen Verteilung der gesteinsgebundenen Gase im Salinar. - Z. geol. Wiss. 17: 353-368. Berlin.

KNIPPING, B. (1984): Der mineralogische und chemische Stoffbestand am Kontakt zwischen Basalten und Evaporitgesteinen der Werra-Serie des deutschen Zechsteins. - Diplomarbeit, Georg-August-Universität, Göttingen, unpub.

KNIPPING, B. (1986): $^{34}S/^{32}S$ ratios of native sulphur in Zechstein 1 evaporites. - Die Naturwissenschaften, 73: 614. Heidelberg.

KNIPPING, B. (1987a): Basaltintrusionen in Zechstein 1-Evaporiten (Werra-Lagerstättenbezirk). - Dissertation. Göttingen.

KNIPPING, B. (1987b): Tertiäre Basalte in Perm-Evaporiten (Z1). - Fortschritte der Mineralogie 65, Bh. 1: 96. Stuttgart.

KNIPPING, B. (1988): Die genetische Interpretation von Salzgesteinskörpern mit Hilfe von Stoffbilanzrechnungen. - Fortschr. Miner. 66, Bh. 1: 82. Stuttgart.

KNIPPING, B. (1989): Basalt intrusions in evaporites. - Lecture notes in earth sciences, vol. 24: pp. 132, Springer, Berlin-Heidelberg-New York-Tokyo.

KNIPPING, B. (1991): Basaltische Gesteine in Zechsteinevaporiten. - Z. Geol. Paläont., Vol. 4, Part 1: 1149-1163. Hannover.

Knipping, B., Herrmann, A.G. (1985): Mineralreaktionen und Stofftransporte an einem Kontakt Basalt - Carnallitit im Kalisalzhorizont Thüringen der Werra-Serie des Zechsteins. - Kali u. Steinsalz 9: 111-124. Essen.

Knissel, W., Fornefeld, M. (1988): Bergmännische Herstellung von Hohlräumen in neuen Entsorgungsbergwerken im Salinar. - In: Vol. 7: 255-273; Deponieren von Abfällen. Clausthaler Kursus zur Umwelttechnik. 17.10. -20.10.1988 in St. Andreasberg. CUTEC, Forschungsverbund Umwelttechnik der TU Clausthal.

Kockert, W. (1969): Zum Problem des $CaCl_2$ in natürlichen Wässern und Salzlösungen des Zechsteins. - Bergakademie 7: 401-403.

Koczy, F. (1939): Heliumbestimmungen an Steinsalz und Sylvin. - Sitzungsberichte der Math.-Nat. Kl., Abt. IIa, Akademie d. Wiss. in Wien: 89-105. Wien.

König, H., Wänke, H., Bien, G.S., Rakestraw, N.W., Suess, H.E. (1964): Helium, neon and argon in the oceans. - Deep-Sea Res. 11: 243-247.

Kowalow, D.W. (1981): Zur Frage des Mechanismus von gasdynamischen Erscheinungen, die sich in Kaliwerken aus dem unmittelbaren Hangenden entwickeln. - Neue Bergbautechnik 11: 99-103. Berlin.

Kramer, J. R. (1965): History of sea water. Constant temperature-pressure equilibrium models compared to liquid inclusion analysis. - Geochim. Cosmochim. Acta 29: 921-945.

Krasnokutski, N.P. (1960): Auftreten von Gas in Kaligruben. - Cited after Gimm and Eckart (1968).

Krause, W.B. (1983): Avery Island brine migration tests: instellation, operation, data collection, and analysis. - Technical Report, pp. 91, ONWI-190 (4), Office of Nuclear Waste Isolation, Battelle Memorial Institute. Columbus, Ohio.

Krejci-Graf, K. (1962): Über Bituminierung und Erdölentstehung. - Freiberger Forschungshefte, C 123: 5-34. Berlin.

Krejci-Graf, K. (1978): Data on the Geochemistry of Oil Field Waters. - Geol. Jb., D 25: 3-174. Hannover.

Kühn, K. (1988): Die untertägige Deponie radioaktiver Abfälle in verschiedenen geologischen Formationen und Gesteinsarten. - In: Vol. 7: 201-207; Deponieren von Abfällen. Clausthaler Kursus zur Umwelttechnik. 17.10. -20.10.1988 in St. Andreasberg. CUTEC, Forschungsverbund Umwelttechnik der TU Clausthal.

Kühn, R. (1951): Nachexkursion im Kaliwerk Hattorf, Philippsthal - als Beitrag zur Kenntnis der Petrographie des Werra-Kaligebietes. - Fortschr. Miner. 29/30: 101-114. Stuttgart.

Kühn, R. (1957): Führung durch das Kalibergwerk Neuhof-Ellers, obere Sohle, nebst einigen Beiträgen zur Petrographie des Werra-Fulda-Kalireviers. - Fortschr. Miner. 35: 60-120. Stuttgart.

Kupfer, D.H. (1963): Structure of salt in Gulf Coast domes. - In: Symposium on Salt: 104-123, ed. by A.C. Bersticker, K.E. Hoekstra, J.F. Hall; The Northern Ohio Geological Society. Cleveland, Ohio.

Kupfer, D.H. (1980): Problems associated with anomalous zones in Louisiana salt stocks, U.S.A.. - In: Fifth Symposium on Salt, Vol. I: 119-134, ed. by A.H. Coogan and L. Hauber: The Northern Ohio Geological Society. Cleveland, Ohio.

Kurze, M., Göring, H.-H. (1964): Ein Beitrag zur Geologie der Stickstoff- und Kohlenwasserstoffvorkommen im Bereich der Sangerhäuser und Mansfelder Mulde. - Freiberger Forschungshefte, A 304: 167-201. Leipzig.

Lang, H.D. (1973): Der Salzstock Hope. Salz. - In: Erläuterungen zu Blatt Schwarmstedt Nr. 3323: 48-49, 66-67. Geologische Karte von Niedersachsen 1:25000. Niedersächsisches Landesamt für Bodenforschung. Hannover.

Langer, M., Wallner, M., Wassmann, Th.H. (1984): Gebirgsmechanische Bearbeitung von Stabilitätsfragen bei Deponiekavernen im Salzgebirge. - Kali u. Steinsalz 9: 66-76. Essen.

Lazar, B., Holland, H.-D. (1988): The analysis of fluid inclusions in halite. - Geochim. Cosmochim. Acta, 52: 485-490.

Leitzke, C., Sroka, A. (1987): Indirekte Überwachung unzugänglicher Speicher- und Deponiekavernen. - Kali u. Steinsalz 9: 334-344. Essen.

LEVY, P.W., LOMAN, J.M., SWYLER, K.J., KLAFFKY, R.W. (1981): Radiation damage studies on synthetic NaCl crystals and natural rock salt for radioactive waste disposal applications. - In: The Technology of High-Level Nuclear Waste Disposal; ed. by P.L. HOFMANN, DOE/TIC-4261, Vol. 1: 136-167, U.S. Dept. of Energy Tech. Info. Div. Oak Ridge, Tennesse.

LIEBSCHER, K.G. (1952): Die Grubengasbekämpfung im Südharz-Kalibergbau. - Bergbautechnik 2: 129-137. Berlin.

LIEDTKE, L., KOPIETZ, J. (1983): Thermomechanical calculations related to thermally induced rock loosening in an underground cavity. - Computers & Structures 17: 891-902.

LITONSKI, A., BIALY, M. (1964): Gasgefahr in den polnischen Salzgruben und Methoden ihrer Bekämpfung. - Cited after GIMM and ECKART (1968).

LÖFFLER, J., SCHULZE, G. (1962): Die Kali- und Steinsalzlagerstätten des Zechsteins in der Deutschen Demokratischen Republik. Teil III: Sachsen-Anhalt. - Freiberger Forschungshefte, C 97/III, pp. 347. Berlin.

LOTZE, F. (1938): Steinsalz und Kalisalze. - pp. 936, Verlag Gebrüder Bornträger, Berlin.

LOTZE, F. (1957): Steinsalz und Kalisalze. I.Teil (Allgemeingeologischer Teil). - pp. 468, Verlag Gebrüder Borntraeger, Berlin.

LUX, K.-H. (1988): Technische Möglichkeiten der Untertage-Deponierung im Salinar. Gebirgsmechanische Aspekte bei der Sonderabfall-Deponierung in soltechnisch aufgefahrenen Kavernen. - In: Vol. 7: 289-317; Deponieren von Abfällen. Clausthaler Kursus zur Umwelttechnik. 17.10. - 20.10.1988 in St. Andreasberg. CUTEC, Forschungsverbund Umwelttechnik der TU Clausthal.

MAASS, I. (1962): Beiträge zur Isotopengeologie der Elemente Wasserstoff, Kohlenstoff und Sauerstoff. - Isotopentechnik 2: 111-116. Leipzig.

MAHTAB, M.A. (1981a): Considerations of gas outbursts in using dome salt mines for storage of oil and nuclear waste. - In: Proc. of the 1st Annual conference on ground control in Mining: 50-58. West Virginia University.

MAHTAB, M.A. (1981b): Occurence and control of gas outbursts in domal salt. - In: 1st Conference on the Mechanical Behavior of Salt, pp. 16, 9.-11. November 1981. The Pennsylvania State University.

MAHLZAHN, E. (1973): Erdöl und Erdgas. - In: Erläuterungen zu Blatt Schwarmstedt Nr.3323: 62-66. Geologische Karte von Niedersachsen 1:25000, Niedersächsisches Landesamt für Bodenforschung. Hannover.

MANAGEMENT OF COMMERCIALLY GENERATED RADIOACTIVE WASTE (1979): Vol 1, US Dept. of Energy, Washington, D.C.

MAYRHOFER, H. (1955): Über ein Langbeinit- und Kainitvorkommen im Ischler Salzgebirge. - Karinthin 30: 94-98.

MAYRHOFER, H. (1973): Der Salzstock Hope. - Cited in: Erläuterungen zu Blatt Schwarmstedt Nr. 3323: 49, 66f.; Geologische Karte von Niedersachsen 1:25000. Niedersächsisches Landesamt für Bodenforschung, Hannover.

MEMMERT, G. (1983): Vorgehensweise bei Sicherheitsanalysen des Endlagers für radioaktive Abfälle. - In: Entsorgung 2: 85-102. Bericht von einer Informationsveranstaltung 23.10.1982 in Hitzacker im Rahmen des Energiedialogs der Bundesregierung; Ed. by: Bundesminister für Forschung und Technologie, Bonn.

MENNING, M. (1986): Zur Dauer des Zechsteins aus magnetostratigraphischer Sicht. - Z. geol. Wiss. 14: 395-404.

MEYER, G.L. (1976): Recent experience with the land burial of solid low-level radioactive wastes. - In: Proc. Symposium of the Management of Radioactive Wastes from the Nuclear Fuel Cycle, Vienna (Austria), March 22-26, 1976; Vol. II: 383-395. IAEA, Vienna.

MIETH, G., FULDA, D., ZEIDLER, W. (1989): Beitrag zum Wirkungsmechanismus von Gasausbrüchen in den Gruben des Südharz-Kalireviers. - Z. geol. Wiss. 17: 407-417. Berlin.

MOORE, J.G., ROGERS, G.C., DOLE, L.R., KESSLER, J.H., MORGAN, M.T., DEVANEY, H.E. (1981): FUETAP concrete - an alternative radioactive waste host. - In: Proc. Intern. Seminar on Chemistry and Process Engineering for High-Level Liquid Waste Solidification, held at Jülich, Germany, June 1-5, Vol. 2: 644-655. Kernforschungsanlage Jülich.

MÜGGE, O. (1928): Über die Entstehung faseriger Minerale und ihre Aggregationsformen. - N. Jb. Min. Geol. Paläont. Beil. A 58: 303-348.

MÜLLER, P., HEYMEL, W. (1956): Verfahren zur Bestimmung der Gaskonzentrationen der Gassalze des Südharz- und Werrakalibergbaues. - Bergbautechnik 6: 313-319. Berlin.

MÜLLER, W. (1958): Über das Auftreten von Kohlensäure im Werra-Kaligebiet. - Freiberger Forschungshefte, A 101, pp. 99. Berlin.

MÜLLER-SCHMITZ, S. (1985): Mineralogisch-petrographische und geochemische Untersuchungen an Salzgesteinen der Staßfurt-, Leine- und Aller-Serie im Salzstock Gorleben (Niedersachsen, BR Deutschland). - Dissertation. Universität Heidelberg, unpub.

MURL : Sonderabfall; Argumente und Informationen, Fakten, Daten, Zahlen. - Der Minister für Umwelt, Raumordnung und Landwirtschaft des Landes Nordrhein-Westfalen.

MURRAY, G.E. (1961): Geology of the Atlantic and Gulf Coastal Province of North America. - pp. 692. Harper & Brothers. New York.

NAGRA AKTUELL (1987): 7, Nr. 3. - Nationale Genossenschaft für die Lagerung radioaktiver Abfälle, Baden, Schweiz.

NAUMANN, M. (1911): Beitrag zur petrographischen Kenntnis der Salzlagerstätte von Glückauf-Sondershausen. - N. Jb. Min. Geol. Paläont., Beilageband XXXII: 578-626. Stuttgart.

NETTEKOVEN, A., GEINITZ, E. (1905): Die Salzlagerstätte von Jessenitz in Mecklenburg. - Mitt. Großherzogl. Mecklenburgischen Geol. L.-A.: 3-17. Rostock.

OCHSENIUS, C. (1877): Die Bildung der Salzlager und ihrer Mutterlaugensalze unter spezieller Berücksichtigung der Flötze von Douglashall in der Egeln'schen Mulde. - pp. 173, Verlag C.E.M. Pfeffer. Halle (Saale).

OELSNER, O. (1961): Ergebnisse neuer Untersuchungen an $CO_2$-führenden Salzen des Werra-Reviers. - Freiberger Forschungshefte, A 183: 5-19. Berlin.

O'NEIL, J.R., JOHNSON, C.M., WHITE, L.D., ROEDDER, E. (1986): The origin of fluids in salt beds of the Delaware basin, New Mexico and Texas. - Applied Geochemistry 1: 265-271.

PANETH, cited after BORN (1934/35 and 1936).

PANETH, F., PETERS, K. (1928a): Heliumuntersuchungen. I. Über eine Methode zum Nachweis kleinster Heliummengen. - Z. physik. Chem. 134: 353-373. Leipzig.

PANETH, F., PETERS, K. (1928b): Heliumuntersuchungen. II. Anwendung des empfindlichen Heliumnachweises auf Fragen der Elementumwandlung. - Z. physik. Chem., Abt. B, Vol. 1: 170-191. Leipzig.

PETERS, H. (1988): Stoffbestand und Genese des Kaliflözes Riedel (K3Ri) im Salzstock Wathlingen-Hänigsen, Werk Niedersachsen-Riedel. - Dissertation. Universität Göttingen, unpub.

PETRICHENKO, O.T. (1973): Methods of study of inclusions in minerals of saline deposits. - pp. 92, Naukova Dumka, Pub. House, Kiew, (in Ukrainian); Translation in: Fluid Inclusion Research, Proceedings of COFFI, Vol. 12, 1979: 214-274. Eds.: E. ROEDDER and A. KOZLOWSKI, The University of Michigan Press.

PHILLIP, W., REINICKE, K.M. (1982): Zur Entstehung und Erschließung der Gasprovinz Osthannover. - Erdoel-Erdgas-Ztschr. 98: 85-90.

PLATE, M. (1988): Entsorgung unter Tage - Chance zur Lösung von Umweltproblemen. - Glückauf 124: 224-229. Essen.

POBORSKI, C., POBORSKI, J. (1964): Über die Untersuchungen in gasführenden Gesteinen im Kujawy-Revier. - Cited after GIMM and ECKART (1968).

POBORSKI, J. (1959): Die Gase in den polnischen Salzgruben. - Cited after GIMM and ECKART (1968).

PRECHT, H. (1879): Die Bestandteile der brennbaren Gase in den Kalisalzbergwerken bei Stassfurt. - Ber. deutsch. Chem. Ges. 12: 557-561. Berlin.

PRECHT, H. (1880): Ueber die Bildung des Wasserstoffs in den Stassfurter Kalisalzbergwerken. - Ber. deutsch. Chem. Gesellschaft 13: 2326-2328. Berlin.

PRECHT, H. (1905): Über die im Kalisalzlager stattgefundene Oxydation des Eisenchlorürs durch Wasserzersetzung unter Bildung von Wasserstoff. - Z. angew. Chem. 18: 1808, 1935f. Berlin.

PTB AKTUELL (1983): Die Schachtvorbohrungen Gorleben 5001 und Gorleben 5002. - Ausgabe 10, Physikalisch-Technische Bundesanstalt, Braunschweig.

REICHARDT, E. (1860): Das Salzbergwerk Stassfurth bei Magdeburg. - Review by L.F. BLEY in: Archiv der Pharmacie, 2. Reihe, Band CIII: 343-350. Hannover.

RICHTER, A. (1962): Die Rotfärbung in den Salzen der deutschen Zechsteinlagerstätten . 1. Teil. - Chemie der Erde, 22: 508-546. Jena.

RICHTER, A. (1964): Die Rotfärbung in den Salzen der deutschen Zechsteinlagerstätten . 2. Teil. - Chemie der Erde, 23: 179-203. Jena.

RICHTER, A., KLARR, K. (1984): Bischofit im Staßfurtflöz der Asse. - Kali und Steinsalz 9: 94-101. Essen.

RICHTER, W. (1953): Untersuchungen zum Auffinden von Kohlensäurenestern im Kalibergbau mit geoelektrischen Meßmethoden. - Freiberger Forschungshefte, C 7: 87-103. Berlin.

RICHTER-BERNBURG, G. (1979): Diskussionsbeitrag. - 191, in: Rede-Gegenrede, Diskussionsprotokolle. Ed. by Deutsches Atomforum e.V.. Bonn.

RIEDEL, L. (1935): Die Fauna des Zechsteins. - In E. FULDA: Zechstein, Handbuch der vergleichenden Stratigraphie Deutschlands: 70-95. Gebrüder Borntraeger, Berlin.

RINGWOOD, A.E. (1978): Safe disposal of high level nuclear reactor wastes: a new strategy. - pp. 64, Canberra, Australia, and Norwalk, Connecticut, USA: Australian National University Press.

RINGWOOD, A.E. (1980): Safe disposal of high-level radioactive wastes. - Fortschr. Miner. 58: 149-168.

RISCHMÜLLER, H. (1972): Salzkavernen zur Speicherung von Rohöl und Erdgas in der Bundesrepublik Deutschland. - Erdoel-Erdgas-Ztschr. 88: 240-248.

ROBERTSON, J.B. (1980): Shallow land burial of low-level radioactive wastes in the USA. - In: Proc. Symposium on Underground disposal of Radioactive Wastes, held at Otaniemi, Finland, July 2-6, 1979; Vol. II: 253-268. IAEA, Vienna.

ROEDDER, E. (1958): Technique for the extraction and partial chemical analysis of fluid-filled inclusions from minerals. - Economic Geology 53: 235-269.

ROEDDER, E. (1984): The fluids in salt. - Am. Mineral. 69: 413-439.

ROTH, H. (1957): Befahrung des Kalisalzbergwerkes "Wintershall" der Gewerkschaft Wintershall in Heringen/Werra. - Fortschr. Miner. 35: 82-88.

RÖTHEMEYER, H. (1981): Konzepte zur Endlagerung in geologischen Formationen. - Proc. Internat. Seminar on Chemistry and Process Engineering for High-Level Liquid Waste Solidification, held at Jülich, Germany, June 1-5, 1985, Vol. 2: 767-783.

RÖTHEMEYER, H. (1988): Beurteilung der Langzeitsicherheit von Endlagern für radioaktive Abfälle. - In: Vol. 7: 319-347; Deponieren von Abfällen. Clausthaler Kursus zur Umwelttechnik. 17.10. - 20.10.1988 in St. Andreasberg. CUTEC, Forschungsverbund Umwelttechnik der TU Clausthal.

RÖTHEMEYER, H., (ED.) (1991): Endlagerung radioaktiver Abfälle. Wegweiser für eine verantwortungsbewußte Entsorgung in der Industriegesellschaft. - pp. 275, VCH Verlagsgesellschaft, Weinheim.

RÖTHEMEYER, H., CLOSS, K.-D. (1981): High-level waste disposal. - Proc. Internat. Conference on World Nuclear Energy-Accomplishments and Perspectives, held at Washington, D.C., November 17-21, 1980, American Nuclear Society 37: 165-175. La Grange Park, Illinois..

ROTHFUCHS, T., WIECZOREK, K., FEDDERSEN, H.K., STAUPENDAHL, G., COYLE, A.J., KALIA, H., ECKERT, J. (1985): Nuclear waste repository simulation experiments. - pp. 225, GSF-Bericht 40/86, T-260, Institut für Tieflagerung. Braunschweig.

SAUNDERS, J., cited after HOY et al. (1962).

SAVCHENKO, V.P. (1958): The formation of free hydrogen in the earth's crust, as determined by the reducing action of the products of radioactive transformations of isotopes. - Geochemistry (GeoKhimiya): 16-25. New York.

SCHAAR, P. (1989): Die Entsorgung von Sonderabfall in Bergbaubetrieben unter Tage aus bergmännischer Sicht. - Glückauf 125: 189-193. Essen.

SCHALLER, cited after OELSNER (1961).

SCHATZEL, ST.J., HYMAN, D.M. (1984): Methane content of Gulf Coast domal rock salt. - Bureau of Mines Report of Investigations 8889, Unites States Department of Interior. Pittsburgh, Pennsylvania.

SCHAUBERGER, O. (1960): Über das Auftreten von Naturgasen im alpinen Salinar. - Erdoel-Zeitschrift 76: 226-233. Wien-Hamburg.

Scheerer (1911): Gasvorkommen in Kalisalzbergwerken. - Zeitschr. für das Berg-, Hütten- und Salinenwesen im Preußischen Staate 59: 212-229. Berlin.

Schleiden, M.J. (1875): Das Salz. Seine Geschichte, seine Symbolik und seine Bedeutung im Menschenleben. - Faksimile-print of the Leipzig-edition (Verlag Engelmann). pp. 278, Verlag Chemie, Weinheim, 1983.

Schmidt, R. (1911): Beschaffenheit und Entstehung parallelfasriger Aggregate von Steinsalz und Gips. - Dissertation, Universität Kiel. Halle (W. Knapp).

Schmidt, R. (1914): Über die Beschaffenheit und Entstehung parallelfasriger Aggregate von Steinsalz und Gips. - Kali 8: 161-166, 197-202, 218-222, 239-245.

Schmiedl, H.-D., Runge, A., Jordan, H., Koch, K., Pilot, J., Elert, K.-H. (1982): Die Deuterium- und Sauerstoff-18-Isotopenanalyse - ein modernes Verfahren zur Bewertung untertägiger Salzlösungsvorkommen in Kali- und Steinsalzgruben. - Z. Geol. Wiss. 10: 73-85.

Schmitt, M. (1987): Isotopen-Geochemische Untersuchungen an Gasspuren aus Salzkernen der Bohrungen Go 1002, Go 1003 und Go 1004 im Salzstock Gorleben. - pp. 35, Firma GCA (Geochemische Analysen), 3160 Lehrte, unpub.

Schober, F., Sroka, A. (1987): Zur Langzeitbelastung über- und untertägiger Anlagen bei Speicher- und Deponiekavernen. - Kali u. Steinsalz 9: 408-414. Essen.

Schoell, M. (1983): Geochemische Untersuchungen an Erdgasen und Kondensaten aus Bohrungen des Salzstocks Gorleben. - pp. 80, Bundesanstalt für Geowissenschaften und Rohstoffe, Archiv-Nr. 94 529, Tagebuch-Nr. 10 537/83, unpub.

Schopper, J.R. (1982): Porosität und Permeabilität. - In: Landoldt-Börnstein, Neue Serie, Band 1: Physical properties of the rocks, part A: 184-303; Springer, Berlin-Heidelberg-New York.

Schott, E. (1989): Zum Aufschluß des Hauptdolomits durch Streckenauffahrungen in der Grube Volkenroda. - Z. geol. Wiss. 17: 401-406. Berlin.

Schrader, R., Ackermann, G., Grund, H. (1960): Entwicklung von Methoden zur Bestimmung des Gasgehaltes in Salzen. - Bergakademie 12: 543-551. Berlin.

Schrader, R., Ackermann, G., Grund, H. (1962): Neue Methoden zur Bestimmung des Gasgehaltes in Salzen, II. Verfahren zur Isolierung der Salzgase für die gaschromatographische Bestimmung. - Acta Chimica 33: 31-38. Budapest.

Schulze, O. (1985): Untersuchung der thermomechanischen Eigenschaften radioaktiv bestrahlter Salzproben. - Abschlußbericht zum F-Vorhaben KWA 51051, Bundesanstalt für Geowissenschaften und Rohstoffe, Hannover.

Schwandt, A. (1973/74): Zusammenhänge zwischen Geologie und Zuflüssen von Salzlösungen und Wässern in Kali - und Steinsalzgrubenfeldern des Saale-Unstrut- und Nordharz- Kaligebietes. - Jb. Geol., Bd. 9, 10: 175-260. Berlin

Slotta, R. (1980): Technische Denkmäler in der Bundesrepublik Deutschland. Die Kali- und Steinsalzindustrie. - pp. 780, Deutsches Bergbau-Museum. Bochum.

Smith, A.G., Briden, J.C. (1977): Mesozoic and Cenozoic paleocontinental maps. - pp. 63, Cambridge University Press. London-New York-Melbourne.

Smits, F., Gentner, W. (1950): Argonbestimmungen an Kalium-Mineralien. I. Bestimmungen an tertiären Kalisalzen. - Geochim. et Cosmochim. Acta 1: 22-27. London.

Snarsky, A. (1963): Bildungsbedingungen der Gaslagerstätten im Thüringer Becken. - Freiberger Forschungshefte, C 165: pp. 42. Leipzig.

Sonderabfall- und Reststoffbestimmungs-Verordnung (1989): - SAbS/Rest-Best.-V., Verordnungsentwurf der Bundesregierung. Stand 03.01.1989. WA II 2-530110/8.

Sonderabfallarten-Katalog. Entwurf 31.01.1989 - Berlin, Umweltbundesamt, Umplist.

Spackeler, G. (1957): Lehrbuch des Kali- und Steinsalzbergbaus. - pp 598, VEB Wilhelm Knapp Verlag, Halle (Saale).

Spies, H. (1985): Erste Ergebnisse einer Abfallbilanz für die Bundesrepublik Deutschland. - Wirtschaft und Statistik 1: 27-34. Kohlhammer, Stuttgart.

SPIES, H. (1987): Ergebnisse einer Abfallbilanz für die Bundesrepublik Deutschland für 1984. - Statistisches Bundesamt, IV E 41.42, Wiesbaden.

STATISTISCHES BUNDESAMT (1987): Öffentliche Abfallbeseitigung 1984. - Fachserie 19, Reihe 1.1.

STÄUBERT, U., STÄUBERT A. (1989): Zum Einfluß der Tektonik auf die Gasführung im Salinar. - Z. geol. Wiss. 17: 369-380. Berlin.

STEIN, C.L. (1985): Preliminary report on fluid inclusions from halites in the Castile and Lower Salado formations of the Delaware basin, southeastern New Mexico. - SAND 83-0451, pp. 42, Sandia National Laboratories. Albuquerque, New Mexico.

STOLLE, E. (1953, 1954): Gasvorkommen in Kalibergwerken des Südharzgebietes. - Bergbautechnik 3: 646-650; 4: 46-52. Berlin.

STOLLE, E. (1974): Salze. - In: Geologie von Thüringen, Eds.: W. HOPPE, G. SEIDEL: 899-912, VEB H. Haack, Gotha-Leipzig.

STONE, R.B. (1987): Underground storage of hazardous waste. - J. Hazardous Materials 14: 23-37. Amsterdam.

STRUTT, R.J. (1908): On helium in saline minerals, and its probable conneciton with potassium. - Proc. of the Roy. Soc. of London, Series A, 81: 278-279. London.

SUTTER, H. (1987): Strategien und Verfahren zur Vermeidung und Verwertung von Sonderabfällen. - In: Sonderabfallentsorgung in Niedersachsen, Dokumentation einer Tagung des Niedersächsischen Umweltministeriums vom 05. - 07.05.1987 in Hannover: 13-38.

SVERDRUP, H.U., JOHNSON, M.W., FLEMING, R.H. (1942): The oceans. - pp. 1087, Prentice-Hall, New York.

TAMMANN, G., SEIDEL, K. (1932): Zur Kenntnis der Kohlensäureausbrüche in Bergwerken. - Z. anorg. u. allgemeine Chem. 205: 209-229.

TÄTIGKEITSBERICHT DER BGR (1985/86): Projektbezogene Forschung an Endlagerstandorten, 2.3.1 Gorleben - 56-64; Bundesanstalt für Geowissenschafen und Rohstoffe, Hannover.

THE DISPOSAL OF RADIOACTIVE WASTE ON LAND (1957). Report of the Committee on Waste Disposal of the Division of Earth Sciences. - pp. 142, National Academy of Sciences - National Research Council, Publication 519, Washington, D.C.

THOMA, K., ECKART, D. (1964): Untersuchungen an gashaltigen Mineralsalzen. I. Teil: Bergmännische Untersuchungen. - Bergakademie 16: 674-676. Berlin.

THOMS, R.L., MARTINEZ, J.D. (1980): Blowouts in domal salt. - In: Fifth Symposium on Salt, Vol. I: 405-411, ed. by A.H. COOGAN and L. HAUBERT: The Northern Ohio Geological Society. Cleveland, Ohio.

TUREKIAN, K.K. (1969): The oceans, streams, and atmosphere. - In: Handbook of Geochemistry, Vol. I: 297-323. Ed. by K.H. WEDEPOHL. Springer, Berlin-Heidelberg-New York.

UNTERIRDISCHE MÜLLDEPONIEN FÜR MASSENABFÄLLE (1988): - In: Baustoffrecycling 2: 38.

VALENTINER, S. (1912): Heliumgehalt im blauen Steinsalz. - Kali 6: 1-3. Halle (Saale).

VAN OPBROEK, G., DEN HARTOG, H.W. (1985): Radiation damage of NaCl: dose rate effects. - J. Phys. C: Solid State Phys. 18: 257-268.

VENZLAFF, H. (1978): Tieflagerung radioaktiver Abfälle aus geologischer Sicht. - Atomwirtschaft-Atomtechnik, 23: 335-338.

VIGIER, G., PANCZUK, T. (1961): Etudes sur les mouvements des banes et recherche des anomalies aux Mines Domaniales de Potasse d'Alsace. - Revue de l'industrie minerale 43: 644-656. St. Etienne.

WALTHER, J.V., ORVILLE, P.M. (1982): Volatile production and transport in regional metamorphism. - Contr. Miner. Petrol. 79: 252-257.

WASSERWIRTSCHAFTLICHE ANFORDERUNGEN AN GESTEINSKAVERNEN ZUM LAGERN WASSERGEFÄHRDENDER STOFFE (ANFORDERUNGSKATALOG). - Gemeinsames Ministerialblatt, Ausgabe B, 40. Jahrgang, Nr. 16: 393-399. Ed. by Bundesminister des Innern, 1989. Bonn.

WEIß, H.M. (1980): Möglichkeiten der Entstehung sowie Art, Umfang und tektonische Stellung von Rissen und Klüften im Salzgebirge. - pp. 96, GSF-Bericht T-200, ISSN 0721-1694. Braunschweig.

WEIZSÄCKER, C.F. VON (1978): Kernenergie. - In: Deutlichkeit - Beiträge zu politischen und religiösen Gegenwartsfragen: 43-72. Verlag Hanser, München-Wien.

WIEDEMANN, H.U. (1988a): Technische Anforderungen an die Abfallablagerung. - In: Behandlung von Sonderabfällen, Vol. 2: 919-982, Ed:. K. J. THOMÉ-KOSMIENSKY, EF-Verlag für Energie- und Umwelttechnik. Berlin.

WIEDEMANN, H.U. (1988b): Untertägige Ablagerung. - In: Behandlung von Sonderabfällen, Vol. 2: 1053-1063, Ed.: K. J. THOMÉ-KOSMIENSKY, EF-Verlag für Energie- und Umwelttechnik. Berlin.

WINTER, U. (1964): Die Anwendung geophysikalischer Verfahren zur Bekämpfung der Gasgefahr im Kalibergbau der DDR. - Bergakademie, 16: 138-145. Berlin.

WOHLENBERG, J. (1982): Dichte der Minerale. - In: Landoldt-Börnstein, Neue Serie, Band 1: Physical properties of the rocks, part A: 66-113. Springer, Berlin-Heidelberg-New York.

WOLF, H. (1965): Zur Aerodynamik der plötzlichen Ausbrüche von Salz und Gas im Werra-Kalibergbau. - Dissertation, Technische Universität in Dresden.

# 21 Subject index

## A

Adolfsglück 96
adsorptive bonding 66
Aller rock salt 72
Aller-Nordstern 92, 96
arid climate 115
Asse 35, 42f
avoidance of wastes 5

## B

backfill 36f
bacterial reduction of sulfates 83
Bartensleben 44, 92ff
basalt magmatism 71, 79f, 83, 87f
110, 116, 120, 122
basalt-carnallitite contact zone 87
Belle Isle 93
Benthe 25
Beresniki 89
Bischofferode 73
Bismarckshall 73
Bleicherode 73
Boyle-Mariotte, law of 89
Braunschweig-Lüneburg 73, 83, 92, 94, 102
Bunde 42

## C

Cambrian rock salt 79
carbon dioxide 66, 68, 80
caverns 24, 60, 63, 119
chemical composition of filter dust 13
chloride type of marine evaporites 62
$CO_2$ 80. See also carbon dioxide
commercial wastes 6
construction wastes 7
crack fillings 137
crack-bound gases 65
crackle carnallite 134, 147
crackle salt 71, 75, 92, 98, 101, 146. See also knister salt
cryogenic fractures 86
cryometry 55

## D

$\delta^{18}O$ 61
$\delta D$ 61
decision-makers 52
decripitation methods 56

deformational events 121
Desdemona 77, 96
Dietlas 73
Dnjepr-Donez depression 74
dynamic metamorphism 82
dynamic nature of geological repository systems 130

## E

earthquakes 116
Eh 61
ERAM 44. See also Bartensleben
Etzel 26, 42
excavated ground 7

## F

federal crude oil reserves 42
fissure fillings 137
fissure-bound gases 65
fluid inclusions 55, 58, 60, 75, 144
formation waters 113, 136
Forsmark 24
fracture fillings 137
fracture-bound gases 92
fracture-filling minerals 164
Frisch-Glück 96

## G

gas explosions 72, 109
gas in cracks 74
gas inclusions 63, 65, 71
gas mixtures 67, 100, 107
gas pressures 74
gas volumes 72, 92, 102
gas-bearing carnallitites 81
geochemical cycles 15ff, 31
geological events 117f
Gulf Coast 71, 92, 95

## H

halokinesis 120
hard-coal-fired plants 13
Hattorf 71
Hauptdolomit 78
Heilbronn 40f, 46, 48ff
Hemelinger carnallite 59
Herfa-Neurode 37, 40f, 45, 48ff
Hessen potash salt seam 59, 71f, 82
high-heat-generating substances 120
Hope 25, 46, 77f, 93ff, 98f
household wastes 6
hydrocarbons 66ff, 77f, 82, 95

hydrofracturing   120
hydrogen   66, 80
hydrogen sulfides   66, 83

**I**

ice age   115
incineration plants   8
incrystalline   65, 74
industrial sludge   46
inflammable gases   108
ingranular   65
intercrystalline   66
intergranular   66
intragranular   65
isotope studies of gases   100

**J**

Jemgum   24, 35, 42, 50

**K**

Kaiseroda   73
knister salt   71, 74f, 92, 98. *See also* crackle
    salt
Kochendorf   42
Konrad iron ore mine   35, 47f
Kupferschiefer   82, 101

**L**

landfills   5, 11, 15f, 22. *See also* surface
    dumps
Leine anhydrite   60
Leine rock salt   60, 72, 74, 97
Leine Valley   93
Leopoldshall   81
limitations of underground repositories   48
liquid hydrocarbons   79, 95
liquid wastes   8
long-term effect   12
long-term impact of nonradioactive residues   14
long-term impact of radioactive residues   14
long-term isolation   16, 18, 35
long-term observations   15
Louisiana   73

**M**

Magdeburg-Halberstadt mining district   67, 81
Marie   44. *See also* ERAM
marine transgression   115
Menzengraben   73
mercury   84
MgSO$_4$ depletion in salt solutions   57

microbiological activity   77
microcracks   85ff
migration   15f, 22, 85, 124, 147
migration velocity   87
mineral-bound gases   66, 92
mixture of solutions   165
mobile components   113, 126
mobility of gases   85
Morsleben   44. *See also* Bartensleben, ERAM
multibarrier geological system   37
Muschelkalk salts   41

**N**

native sulfur   83
nitrogen   66, 82
noble gases   66, 84

**O**

Obrigheim   42
oil   62, 95
organic matter   62, 77ff, 90, 92f, 97, 107
oxygen   66

**P**

paths for mobile components   120, 125
permeability   119
pH   61
plastic deformation   120
polluting effect of wastes   52
polythermal mineral formation   139
porosity   119
processing of wastes   5
public acceptance   48, 50
public interest   52

**Q**

quantities of hazardous wastes   9
quaternary systems   56
quinary system   56, 76

**R**

radiolysis   81, 102
reducing conditions   78
reduction of wastes   5
refuse-fired power plants   13
repository caverns   26
repository chambers   28
reprocessed spent nuclear fuel   12, 20
resevoirs for gas   101
residence times of elements   17
Ronnenberg   59

Rotliegend   24, 82, 84, 119

**S**

Sachsen-Weimar   71, 73f
salt caverns   24, 50
sewage sludge   6, 13
shaft drillings   97f
Siberia   79
social factors   31, 39
Soligorsk   75
Solikamsk   67, 73, 88
solution equilibria   60, 149, 150
solution metamorphism   55, 82, 86, 107
solution mining   20, 24f, 42, 63
source of hazardous wastes   10
Southern Harz mining district   73
Staßfurt carbonate   72, 78f
Staßfurt potash salt seam 59, 67, 73, 99, 125f,
      132, 134, 137, 150, 164
Staßfurt potash seam   45, 73
Staßfurt rock salt   60, 69, 78, 87, 92, 99, 126f,
      133, 138, 144, 149, 166
Stetten   42
Subduction zones   20
subrosion   118
sulfate type of marine evaporites   62
surface dumps   6f. *See also* landfills

**T**

Teutonia   93
Thailand   89
thermal metamorphism   55, 57
thermometry   55
Thiede   45
Thiederhall   45
Thüringen   59
Thüringen potash salt seam   59, 71f, 82
toxicity   6, 12, 14
toxicity indices   12, 14

**U**

unreprocessed spent nuclear fuel   12
Unstrut-Saale mining district   73
upper Kama   75

**V**

Vienenburg   59
voids   93
Volkenroda   78

**W**

Werra carbonate   72
Werra rock salt   71f
Wintershall   37
Wohlverwahrt-Nammen   48f
Wustrow   93f, 99

Printing: Druckhaus Beltz, Hemsbach
Binding: Buchbinderei Schäffer, Grünstadt

# Lecture Notes in Earth Sciences

Vol. 1: Sedimentary and Evolutionary Cycles. Edited by U. Bayer and A. Seilacher. VI, 465 pages. 1985. (out of print).

Vol. 2: U. Bayer, Pattern Recognition Problems in Geology and Paleontology. VII, 229 pages. 1985.

Vol. 3: Th. Aigner, Storm Depositional Systems. VIII, 174 pages. 1985.

Vol. 4: Aspects of Fluvial Sedimentation in the Lower Triassic Buntsandstein of Europe. Edited by D. Mader. VIII, 626 pages. 1985.

Vol. 5: Paleogeothermics. Edited by G. Buntebarth and L. Stegena. II, 234 pages. 1986.

Vol. 6: W. Ricken, Diagenetic Bedding. X, 210 pages. 1986.

Vol. 7: Mathematical and Numerical Techniques in Physical Geodesy. Edited by H. Sünkel. IX, 548 pages. 1986.

Vol. 8: Global Bio-Events. Edited by O. H. Walliser. IX, 442 pages. 1986.

Vol. 9: G. Gerdes, W. E. Krumbein, Biolaminated Deposits. IX, 183 pages. 1987.

Vol. 10: T.M. Peryt (Ed.), The Zechstein Facies in Europe. V, 272 pages. 1987.

Vol. 11: L. Landner (Ed.), Contamination of the Environment. Proceedings, 1986. VII, 190 pages.1987.

Vol. 12: S. Turner (Ed.), Applied Geodesy. VIII, 393 pages. 1987.

Vol. 13: T. M. Peryt (Ed.), Evaporite Basins. V, 188 pages. 1987.

Vol. 14: N. Cristescu, H. I. Ene (Eds.), Rock and Soil Rheology. VIII, 289 pages. 1988.

Vol. 15: V. H. Jacobshagen (Ed.), The Atlas System of Morocco. VI, 499 pages. 1988.

Vol. 16: H. Wanner, U. Siegenthaler (Eds.), Long and Short Term Variability of Climate. VII, 175 pages. 1988.

Vol. 17: H. Bahlburg, Ch. Breitkreuz, P. Giese (Eds.), The Southern Central Andes. VIII, 261 pages. 1988.

Vol. 18: N.M.S. Rock, Numerical Geology. XI, 427 pages. 1988.

Vol. 19: E. Groten, R. Strauß (Eds.), GPS-Techniques Applied to Geodesy and Surveying. XVII, 532 pages. 1988.

Vol. 20: P. Baccini (Ed.), The Landfill. IX, 439 pages. 1989.

Vol. 21: U. Förstner, Contaminated Sediments. V, 157 pages. 1989.

Vol. 22: I. I. Mueller, S. Zerbini (Eds.), The Interdisciplinary Role of Space Geodesy. XV, 300 pages. 1989.

Vol. 23: K. B. Föllmi, Evolution of the Mid-Cretaceous Triad. VII, 153 pages. 1989.

Vol. 24: B. Knipping, Basalt Intrusions in Evaporites. VI, 132 pages. 1989.

Vol. 25: F. Sansò, R. Rummel (Eds.), Theory of Satellite Geodesy and Gravity Field Theory. XII, 491 pages. 1989.

Vol. 26: R. D. Stoll, Sediment Acoustics. V, 155 pages. 1989.

Vol. 27: G.-P. Merkler, H. Militzer, H. Hötzl, H. Armbruster, J. Brauns (Eds.), Detection of Subsurface Flow Phenomena. IX, 514 pages. 1989.

Vol. 28: V. Mosbrugger, The Tree Habit in Land Plants. V, 161 pages. 1990.

Vol. 29: F. K. Brunner, C. Rizos (Eds.), Developments in Four-Dimensional Geodesy. X, 264 pages. 1990.

Vol. 30: E. G. Kauffman, O.H. Walliser (Eds.), Extinction Events in Earth History. VI, 432 pages. 1990.

Vol. 31: K.-R. Koch, Bayesian Inference with Geodetic Applications. IX, 198 pages. 1990.

Vol. 32: B. Lehmann, Metallogeny of Tin. VIII, 211 pages. 1990.

Vol. 33: B. Allard, H. Borén, A. Grimvall (Eds.), Humic Substances in the Aquatic and Terrestrial Environment. VIII, 514 pages. 1991.

Vol. 34: R. Stein, Accumulation of Organic Carbon in Marine Sediments. XIII, 217 pages. 1991.

Vol. 35: L. Håkanson, Ecometric and Dynamic Modelling. VI, 158 pages. 1991.

Vol. 36: D. Shangguan, Cellular Growth of Crystals. XV, 209 pages. 1991.

Vol. 37: A. Armanini, G. Di Silvio (Eds.), Fluvial Hydraulics of Mountain Regions. X, 468 pages. 1991.

Vol. 38: W. Smykatz-Kloss, S. St. J. Warne, Thermal Analysis in the Geosciences. XII, 379 pages. 1991.

Vol. 39: S.-E. Hjelt, Pragmatic Inversion of Geophysical Data. IX, 262 pages. 1992.

Vol. 40: S. W. Petters, Regional Geology of Africa. XXIII, 722 pages. 1991.

Vol. 41: R. Pflug, J. W. Harbaugh (Eds.), Computer Graphics in Geology. XVII, 298 pages. 1992.

Vol. 42: A. Cendrero, G. Lüttig, F. Chr. Wolff (Eds.), Planning the Use of the Earth's Surface. IX, 556 pages. 1992.

Vol. 43: N. Clauer, S. Chaudhuri (Eds.), Isotopic Signatures and Sedimentary Records. VIII, 529 pages. 1992.

Vol. 45: A. G. Herrmann, B. Knipping, Waste Disposal and Evaporites. XII, 193 pages. 1993.